BERNARD STIEGLER

Also Available from Bloomsbury

The Re-Enchantment of the World, Bernard Stiegler
Aesthetics, Digital Studies and Bernard Stiegler, ed. Noel Fitzpatrick, Néill O'Dwyer, Michael O'Hara

BERNARD STIEGLER

Memories of the Future

Bart Buseyne, Georgios Tsagdis and Paul Willemarck

BLOOMSBURY ACADEMIC
LONDON • NEW YORK • OXFORD • NEW DELHI • SYDNEY

BLOOMSBURY ACADEMIC
Bloomsbury Publishing Plc, 50 Bedford Square, London, WC1B 3DP, UK
Bloomsbury Publishing Inc, 1385 Broadway, New York, NY 10018, USA
Bloomsbury Publishing Ireland, 29 Earlsfort Terrace, Dublin 2, D02 AY28, Ireland

BLOOMSBURY, BLOOMSBURY ACADEMIC and the Diana logo
are trademarks of Bloomsbury Publishing Plc

First published in Great Britain 2024
This paperback edition published 2025

Copyright © Bart Buseyne, Georgios Tsagdis and Paul Willemarck, and Contributors, 2024

Bart Buseyne, Georgios Tsagdis and Paul Willemarck have asserted their right under the
Copyright, Designs and Patents Act, 1988, to be identified as Author of this work.

For legal purposes the Acknowledgements on p. xii constitute
an extension of this copyright page.

Series design by Charlotte Daniels
Cover image: Attic red-figure pottery krater depicting a mythological scene
(AN1896-1908.G.275). (© Ashmolean Museum, University of Oxford)

All rights reserved. No part of this publication may be: i) reproduced or transmitted in any form, electronic or mechanical, including photocopying, recording or by means of any information storage or retrieval system without prior permission in writing from the publishers; or ii) used or reproduced in any way for the training, development or operation of artificial intelligence (AI) technologies, including generative AI technologies. The rights holders expressly reserve this publication from the text and data mining exception as per Article 4(3) of the Digital Single Market Directive (EU) 2019/790.

Bloomsbury Publishing Inc does not have any control over, or responsibility for, any third-party websites referred to or in this book. All internet addresses given in this book were correct at the time of going to press. The author and publisher regret any inconvenience caused if addresses have changed or sites have ceased to exist, but can accept no responsibility for any such changes.

A catalogue record for this book is available from the British Library.

A catalog record for this book is available from the Library of Congress.

ISBN: HB: 978-1-3504-1044-2
PB: 978-1-3504-1048-0
ePDF: 978-1-3504-1045-9
eBook: 978-1-3504-1046-6

Typeset by Integra Software Services Pvt. Ltd.

For product safety related questions contact productsafety@bloomsbury.com.

To find out more about our authors and books visit www.bloomsbury.com
and sign up for our newsletters.

CONTENTS

Contributors   viii
Acknowledgements   xii

Chapter 1
INTRODUCTION: THE INVITATION OF MEMORY   1
   Bart Buseyne, Georgios Tsagdis & Paul Willemarck

Part I
IN MEMORIAM: BERNARD STIEGLER

Chapter 2
NOBLE NEGANTHROPOLOGIST: REMEMBERING BERNARD STIEGLER
(1 APRIL 1952–5 AUGUST 2020)   17
   Pieter Lemmens

Chapter 3
STIEGLER, MELANCHOLY, NEGATIVITY (FUNERAL SONG FOR
BERNARD)   27
   Jean-Luc Nancy

Chapter 4
JUST THIS, WRITTEN JUST HERE AND JUST NOW, BY JUST THIS
INDIVIDUAL IN JUST THIS MOOD   35
   Daniel Ross

Part II
INTERMITTENCE: CARING TO BELIEVE

Chapter 5
CARE AS INVENTION: A TRIBUTE TO BERNARD STIEGLER   53
   Anaïs Nony

Chapter 6
AGAINST SIMPLIFICATION: THE INTERMITTENCE OF LIFE   63
   Gerald Moore

Chapter 7
STIEGLER'S HANDS: TERTIARY RETENTIONS AND THE BELIEF
OF REASON 77
    Paul Willemarck

Part III
THINKING *DIFFÉRANCE*: LIFE, TECHNICS, EPOCHALITY

Chapter 8
NEGENTROPY AND *DIFFÉRANCE*: STIEGLER'S MEMORIES OF
THE FUTURE 97
    Georgios Tsagdis

Chapter 9
WHERE THERE IS NO WORLD AND NO EPOCH: BERNARD STIEGLER'S
THINKING OF THE ENTROPOCENE 107
    Erich Hörl

Chapter 10
DIFFÉRANCE AND EPOCHALITY: STIEGLER'S *TOURS* 125
    Donovan Stewart

Part IV
CREATIVE ORGANOLOGIES: WORKS OF INVENTION

Chapter 11
PHILOSOPHY THROUGH ACTING 139
    Bart Buseyne

Chapter 12
TAKING CARE OF DIGITAL TECHNOLOGIES 155
    Vincent Puig

Chapter 13
PLAINE COMMUNE, CONTRIBUTIVE LEARNING TERRITORY 167
    Maël Montévil

Chapter 14
TOWARDS A BIFURCATION: INTERNATION AND INTERSCIENCE IN
THE TWENTY-FIRST CENTURY 175
    Anne Alombert

Chapter 15
A *SCHOLĒ* FOR THE THUNBERG GENERATION 183
   Victor Chaix

Chapter 16
ANOTHER SOCIAL NETWORK IS POSSIBLE! 191
   Harry Halpin & Geert Lovink

Part V
ECHOES: INDIVIDUATING ART

Chapter 17
MNEMOTECHNICS, ECHO AND THE DISCRETE VOICE 205
   Mischa Twitchin

Chapter 18
BERNARD STIEGLER'S LOVE OF MUSIC 217
   Susanna Lindberg

Part VI
AN UNFINISHED CONVERSATION

Chapter 19
ONTOLOGICAL DIFFERENCE, TECHNOLOGICAL *DIFFÉRANCE*
AND SEMANTIC DIFFERENCE: THE PROBLEM OF A DECENTRED
RECONSTRUCTION OF PHILOSOPHY AFTER 'DECONSTRUCTION' 233
   Jean-Hugues Barthélémy

Index 241

CONTRIBUTORS

Anne Alombert is teacher and researcher in philosophy at the Catholic University of Lille. She is the author of a thesis in philosophy which focuses on the relationships between life, technics and spirit in the work of Gilbert Simondon and Jacques Derrida. Her researches focus on the works of Simondon, Derrida and Stiegler, particularly on the epistemological and anthropological issues raised by technics and digital technologies. She participated in the development of the contributory research programme *Plaine Commune Territoire Apprenant Contributif* directed by Bernard Stiegler from 2017 to 2020. She is the co-author of the book *Bifurcate*, written with the Internation collective.

Jean-Hugues Barthélémy is a philosopher and associated researcher at the Paris-Nanterre University. After having published several books of reference on the work of Gilbert Simondon (2005; 2008; 2014; 2015), acted as the chief editor of the *Cahiers Simondon* (2009–15) and as the director of the Centre international des études simondoniennes at the Maison des Sciences de l'Homme Paris-Nord (2014–19), he has brought out *La Société de l'invention: Pour une architectonique philosophique de l'âge écologique* (Paris: Éditions Matériologiques, 2018), *Ego Alter: Dialogues pour l'avenir de la Terre* (Paris: Éditions Matériologiques, 2021), and *Manifeste pour l'écologie humaine* (Paris: Actes Sud, 2022). His book *La Philosophie du paradoxe: Prolégomènes à la Relativité philosophique* (Paris: Éditions Matériologiques, 2023) is forthcoming.

Bart Buseyne studied philosophy and applied philosophy at the universities of Leuven, Hull, and Paris I. He translated several essays by Stiegler into Dutch, mostly in collaboration with Judith Wambacq, with whom he also interviewed Stiegler for *De Uil van Minerva*. He is affiliated to KBR, the Royal Library of Belgium.

Victor Chaix is vice-president of the Association des Amis de la Génération Thunberg – Ars Industrialis (AAGT-AI), of which he is a founding member with Bernard Stiegler, Giuseppe Longo and other scientists or activists. He has had multiple experiences in journalism, notably Le Monde and Reporterre in France, and has also involved himself in the climate movement Extinction Rebellion (XR) – mainly as a 'mediactivist'. After having worked for a year at the Institute of Research and Innovation in Paris, he now pursues his studies at the Università di Bologna, in the Digital Humanities and Digital Knowledge (DHDK) programme.

Harry Halpin is a Research Affiliate at the Center Leo Apostel at Vrije Universiteit Brussel and co-founder of Nym Technologies, a privacy start-up. Previously, he worked as project coordinator of the European Commission NEXTLEAP project at Inria (Paris) and finished his postdoctoral studies under Bernard Stiegler at IRI. He received a Ph.D. from the University of Edinburgh supervised by Andy Clark and worked with Tim Berners-Lee at the World Wide Web Consortium at MIT. He has published over fifty articles in computer science and philosophy.

Erich Hörl is Professor of Media Culture and Media Philosophy at the Institute of Culture and Aesthetics of Digital Media (ICAM), Leuphana University of Lüneburg. He is also vice-president of Leuphana. His main research interests concern the development of a general ecology of media and technologies as well as a critique of Environmentality as the current form of power/knowledge. He publishes widely on the theoretical challenges and the historical becoming of today's technological condition.

Pieter Lemmens teaches philosophy and ethics at the Institute for Science in Society (ISiS) at the Radboud University in Nijmegen, the Netherlands. He has published on themes in the philosophy of technology, on the work of Martin Heidegger, Peter Sloterdijk and Bernard Stiegler, on post-autonomist Marxism and on themes in philosophical anthropology and (post)phenomenology. Current interests are the political and ecological potentials of new digital media and philosophy of technology in the age of the Anthropocene.

Susanna Lindberg is professor of continental philosophy at Leiden University, the Netherlands. She is a specialist of German idealism, phenomenology and contemporary French philosophy. In recent years, her research has carried on the question of technology. Her publications include *Techniques en philosophie* (Hermann, 2020), *Le monde défait: L'être au monde aujourd'hui* (Hermann, 2016), *Heidegger contre Hegel: Les irréconciliables* and *Entre Heidegger et Hegel: L'éclosion et vie de l'être* (L'Harmattan, 2010). She has edited several collected volumes, notably *The Ethos of Digital Environments. Technology, Literary Theory and Philosophy* (with Hana Roine, forthcoming at Routledge, 2021), *The End of the World* (with Marcia Sá Cavalcante Schuback, Rowman and Littlefield, 2017) and *Europe Beyond Universalism and Particularism* (with Sergei Prozorov & Mika Ojakangas, Palgrave Macmillan, 2014). In addition to this, she has published a number of academic articles.

Geert Lovink is a Dutch media theorist, internet critic and founder of the Institute of Network Cultures (www.networkcultures.org). He is the author of *Social Media Abyss* (2016), *Sad by Design* (2019) and *Stuck on the Platform* (2022).

Gerald Moore is professor of French Philosophy and Digital Studies and Associate Director of the Centre for Culture and Ecology at Durham University, UK. He is also a director of the Association des Amis de la Génération Thunberg and, following Stiegler's death, academic director of the Institute for Research and Innovation, Paris. He is co-editor (with Christina Howells) of *Stiegler and Technics*

(Edinburgh University Press, 2013) and co-author (with Bernard Stiegler, Martin Crowley and Ian James) of *Learning to Live Again* (Polity, 2022). His forthcoming monograph, *Anthropocene Animals*, is the product of a decade of close collaboration with Stiegler.

Maël Montévil is a theoretical biologist working at the crossroads of experimental biology, mathematics and philosophy. His work focuses on the theoretical foundations of biology and the role that mathematics can play in this field. He also addresses current issues such as endocrine disruptors and, more generally, the Anthropocene. Maël Montévil started working at the Institute for Research and Innovation (IRI) with Bernard Stiegler in January 2018. He works also at the Institut de Philosophie des Sciences et des Techniques, Université Paris 1, Panthéon-Sorbonne, with a grant of the Cogito Foundation. Most of his publications are available at https://montevil.org.

Jean-Luc Nancy was Professor Emeritus of Philosophy at the University of Strasbourg. Among his many books are *The Inoperative Community* (1991), *The Sense of the World* (1998), *A Finite Thinking* (2003), *The Pleasure in Drawing* (2013) and more recently *An All-Too-Human-Virus* (2020).

Anaïs Nony (Ph.D., University of Minnesota) is Lecturer in the Department of French at University College Cork and Research Associate at the Center for the Study of Race, Gender & Class at the University of Johannesburg. She researches and teaches media art and philosophy of technique, with a focus on gender, class and race in relation to technologies.

Vincent Puig has been studying and developing new technologies at IRCAM and Centre Pompidou since 1993. In 2006, he co-founded with Bernard Stiegler the Institute for Research and Innovation (IRI) which develops contributive research addressing epistemological, cultural, social and economic issues both in the framework of the *Digital Studies* network set up in 2012 and in the Contributive Learning Territory initiated in May 2016 in Seine-Saint-Denis. Both dimensions are presented in the collective book *Bifurquer*, the last directed by Bernard Stiegler, to which Vincent Puig contributed. He is currently scientific delegate of the Digital and citizenship Chair of the Catholic University of Paris, adviser for IMERA (Aix-Marseille University), Cap Digital and City councillor in the Paris Region.

Daniel Ross is the author of *Psychopolitical Anaphylaxis: Steps Towards a Metacosmics* (Open Humanities Press, 2021) and *Violent Democracy* (Cambridge University Press, 2004), and has translated eleven books by Bernard Stiegler into English, most recently *Nanjing Lectures 2016–2019* (Open Humanities Press, 2020). He is also the co-director of the film *The Ister*, which premiered at the 2004 International Film Festival Rotterdam, and won prizes in Montreal and Marseille. It was through making that film that his association with Stiegler began.

Donovan Stewart is a doctoral candidate at the Centre for Continental Philosophy of the Leiden University (the Netherlands). He received a Research M.A. and M.A. in Philosophy from KU Leuven (Belgium), and a B.A. from Bard College Berlin (Germany). His work draws from twentieth and twenty-first-century French thought, with a current focus on the relation between violence and justice, orientated by the themes of ecology, technique and nihilism.

Georgios Tsagdis is postdoctoral researcher in philosophy of technology at Wageningen University and Research. His essays have been published in various collections and international journals, among which are *Parallax*, *Philosophy Today* and *Studia Phaenomenologica*. His editorials include the special issues 'Intersections: at the Technophysics of Space' (*Azimuth*, 2017), 'Of Times: Arrested, Resigned, Imagined' (*International Journal of Philosophical Studies*, 2020) and the collective volume *Derrida's Politics of Friendship: Amity and Enmity* (Edinburgh University Press, 2022). He is the founder of the continental philosophy network *Minor Torus*.

Mischa Twitchin is a lecturer in the Theatre and Performance Dept., at Goldsmiths, University of London: https://www.gold.ac.uk/theatre-performance/staff/twitchin-dr-mischa/. He was a British Academy Post-doctoral Fellow (2014–17) and has contributed chapters to several collected volumes, as well as articles in journals such as *Performance Research* (an issue of which, 'On Animism', he also co-edited). His book, *The Theatre of Death – the Uncanny in Mimesis: Tadeusz Kantor, Aby Warburg and an Iconology of the Actor*, was published by Palgrave Macmillan in their Performance Philosophy series; and examples of his own performance- and essay-films can be seen on Vimeo: http://vimeo.com/user13124826/videos.

Paul Willemarck is an independent researcher working in France. He is Co-founder of Junction Phenomenology (rudolfboehm.org)

ACKNOWLEDGEMENTS

We are grateful to our colleagues and co-organizers of the symposium *Memory for the Future: Thinking with Bernard Stiegler* that took place on the 3rd and 4th of December 2020 and constituted the inception for the present collection. Particular thanks are due to Susanna Lindberg of the Leiden Centre of Continental Philosophy (Leiden University) and Pieter Lemmens of the Institute for Science in Society (Radboud University Nijmegen) for paving the institutional way of the event. For the wealth of our exchanges during and around those two days, we salute all the event's participants, and in particular Erik Bordeleau, Jan Masschelein, Sjoerd van Tuinen and Judith Wambacq. We would like, in turn, to acknowledge the authors of this volume, not only for their invaluable contributions that carry forth the work of Bernard Stiegler, but also for their accommodating attitude and deep commitment to this project. We wish, further, to extend our gratitude to Howard Caygill, Arne De Boever and Erich Hörl, for their advice and support along the way. Finally, a word of thanks goes to Liza Thompson and Katrina Calsado, from Bloomsbury Academic, for guiding us through the home stretch.

Bart Buseyne, Georgios Tsagdis, Paul Willemarck

Chapter 1

INTRODUCTION: THE INVITATION OF MEMORY

Bart Buseyne, Georgios Tsagdis & Paul Willemarck

In the thick of yet another of the century's crises, bio-technological as much as bio-political, the passing of Bernard Stiegler clouded into a darker shade of grey our overcast skies. As an intellectual, Stiegler had never been more present in the *agora*. His contributions were felt increasingly stronger among non-specialist circles, beginning to influence domains far beyond the academic, artistic and experimental techno-scientific enclaves where he had been originally received. At the moment of his passing, Stiegler represented for many of us a hope.

His sharp sensibility for signs of change in the public sphere, this seismographic acuteness to register and to elucidate the symptomatic significance of events that for most of us passed as anecdotal, lent particular gravity to the interventions he initiated. We recall, for instance, his interpretation of the Nanterre massacre in 2002, in which Richard Durn was led to the murder of eight councillors and the injury of nineteen others, before taking his own life. Stiegler saw in Durn's explicit cry for existence a 'structural privation of his primordial narcissistic capacities',[1] a loss of 'the power to signify'[2] and an extreme expression of the evermore common loss of individuation.[3] Similarly, Stiegler was deeply affected by and subsequently drew attention to the testimony of the fifteen-year-old Florian, who in 2015 spoke of his generation's complete loss of dreams and ideals, paralysed by the fear of being the last, or among the last of generations.[4]

In all this, Stiegler was personally implicated. Philosophy was not for him a profession, an exercise, mere intellectual curiosity. As he confessed in 'How I became a philosopher',[5] he found himself in a singular relationship to the practice of philosophy, a relationship informed by the gravity of the accident and the import of agency: the necessity of the latter proceeding upon the default of the former. His theorization of the accidental formed a vantage point to access the historical foreclosed in Heidegger's notion of the 'world-historical' (*das Weltgeschichtliche*), as much as in Frankfurt School's critique of capitalism. Such access enabled him to pass into act. Among numerous interventions, his address to the Secretary-General of the United Nations in November 2019 formed a high point, as he urged for action in the name of life and in the spirit of the young generation, set in sharp relief by the movement *Youth for Climate*.[6]

His suicide only a few months later came thus as an abrupt anti-climax of a verdict on older generations that seemed to be gaining traction. As the call for accountability, if not responsibility, addressed by the youth to the generations that precipitated the environmental catastrophe was growing louder, the world at large and Stiegler's life in specific, came simultaneously to a halt. His failing health was refracted into the ailing of the globe and amplified by the violent idiocy of bio-techno-political apparatuses, who summed up their responses in a series of prohibitions, controls and mandates, barely capable of fathoming the complexity of the crisis they had engendered, a crisis long foreseen and foretold.

His loss of hope is our loss. But this loss marks also a crossroad; Stiegler's passing is a passage. Under the sign of this suicide we find the call of continuing his work, which is not only inscribed in countless tertiary retentions, but which also inhabits a great host of thinkers and practitioners across the globe. In fidelity to this call, the Leiden University Centre for Continental Philosophy (LCCP) and the Institute for Science in Society (ISiS) of the Radboud University of Nijmegen organized on the 3rd and 4th of December 2020 the symposium: *Memory for the Future: Thinking with Bernard Stiegler*. The present volume collates the thoughts we shared on the occasion, here further elaborated and developed. To the memory of Stiegler and to the memory of our symposium in his memory, this book offers yet another memory, inviting many more to come. Moreover, 'Memories of the Future' in its slight prepositional reformulation as the subtitle of this collection reissues the title of an exhibition which Stiegler organized at the Centre Pompidou in Paris ('Mémoires du futur, bibliothèques et technologies', 21 October 1987–18 January 1988); this exhibition, which involved also a colloquium (3–5 December 1987), with the participation of luminaries such as Derrida and Lyotard, raised very early the conglomerate of questions that today constitute the core of Stiegler's legacy.[7]

In the remainder of this introduction, we offer an overview of what is to follow, arranged in six vignettes, six sections, that engage with different facets of Stiegler's thought, life and work and their influence upon various fields of practice and modes of enquiry. We hope that you will welcome this invitation to a thinking (*penser*) that cares (*panser*).

In memoriam: Bernard Stiegler

The first section of this book gathers a few texts that evoke what kind of person of Bernard Stiegler was. Stiegler was more than a philosopher. He was a special witness of our time, a generous listener, a courageous speaker, but also a melancholy dreamer who knew there was no easy way out of the situation we live in. It seems he had a special gift for understanding differently the scope of well-known characteristics of our time such as technology, capitalism or the devastating consequences of human activity. He registered every new opinion and tried to respond to it without delay. He was a relentless worker who firmly believed in the necessity of *parrhesia* and in the value of our historical experience. Pieter Lemmens's *In memoriam* relates a few episodes of this trajectory that show his intellectual generosity. He shows how thorough Stiegler was about discussing the work of his contemporaries and how much effort he put into listening to the younger.

Stiegler set out to think the question of technics anew in a time when technics had obviously become the motor of change in society. The big divide between *epistēmē* and *technē* characterizes philosophy from the start up to Heidegger or the Frankfurt School.[8] Using the work of Leroi-Gourhan, Bertrand Gille and Gilbert Simondon, Stiegler questioned Heidegger's claim that the essence of technics is not technical.[9] Pieter Lemmens shows how Stiegler conceived the role of technics as a pharmacological trauma, the denial of which would eventually result in a proletarianization of thought itself. Against philosophy's customary repression of the role of technics into mere instrumentality, Stiegler articulated a new conception of recollection to wage war against the denoetization of the mind. Today reticular capitalism has taken hold of attention in order to redirect it for its own profit. As a consequence, desire is being reduced to pulsion, in an evolution that is threatening the project of society as a whole. But the ignorance of the role and power of technics is even more endangering for life itself.

The development of thermodynamics established an awareness of the subjection of all living organisms to an entropic becoming, that only open systems can partially counteract. Daniel Ross iterates the late readings of Stiegler as signposts along a spellbound history running to its collapse. From Ludwig Boltzmann, to Vladimir Vernadsky, Alfred Lotka and Arnold Toynbee, the insight into the entropic condition of life seems to allow no escape from its fatal outcome. In response to it and against the transhumanist illusions of bio-technological immortality, Stiegler devoted himself to the possibility of creating anti-entropic processes.[10] For Stiegler, this task, which he termed a neganthropology and to which he would devote the last years of his life,[11] is particularly pressing, in view of the human impact on the earth's becoming. Ross draws the lines of the huge task of taking care of the biosphere in respect to entropy and the improbable chance of our collective capacity to tackle it. With Stiegler and Simondon, he envisions the possibility of a second birth in *transindividuation*.

Three motives come together in Stiegler's work: the incapacity or insufficiency of the necromass we inherit, the critique of proletarianization and the question of entropy. All three of them are staged by Stiegler as problematic aspects of the technical condition of the mortals that we are. They figure weaknesses, regressive tendencies, which often express themselves in violence or destruction. Stiegler understands the rise of philosophy in ancient Greece as that of the question of how it is possible for mortals to live together. It is a question of being sensible (*aidos* and *dikē*) to the requirement of a measure, one that is always already lost in the experience we inherit. As such, the measure is always to come: the condition of humanity is one of default, implying the necessity of a leap in view of our melancholy desire for perfection.

Closing this section Jean-Luc Nancy examines the heritage of our humanity, by engaging in a meditation on Stiegler's notion of the necessary default. To Stiegler, the necessary default is the default of an assignable origin in humans.[12] This experience of a lack of origin is not merely transformative; it is, rather, constitutive. Melancholy is, in turn, the default sentiment, a sentiment by default, of ourselves as mortal beings. It is the sentiment of our historicity, which is our access to the necromass of what we inherit. Melancholy turns the task of caring into a dream or a promise. It can only be attained by a leap into 'that from which no sense can arise', since it will only be accessible as the necromass that it will have become. Nancy evokes Hegel's tarrying of the spirit in death. He marks Stiegler's indecision about the negativity of the condition of humans and the necessity of a halt.

Intermittence: Caring to believe

The figure of the human remains of primary significance throughout Stiegler's work. In the *ad hominem* of his philosophical practice and his 'acting out', the history of hominization and the promise of humanity meet. This section stages this encounter by bringing together three essays that thematize the conditions of human life in general, and its foundational intermittence, while observing closely the human life that took place under the name of Bernard Stiegler.

Here, care is foundational. As Anaïs Nony showcases, care connects the 'retentional trace of the past (the *pansable*)' with 'the protentional field of the future (the *suspensible*)', transforming thus the *impansé* (uncured) into the *pensable* (what can be thought). Being constituted by techniques of care, thought (*nous*) invents the passage from knowledge to praxis. This passage is threatened by the technological short-circuiting and usurpation of *nous* by computational processes of control which obliterate the attentional forms of both psychic and collective care.[13]

Nonetheless, Stiegler thematizes technique as the originary human (non-inhuman) supplement. As Nony observes poignantly, human noetic life becomes technical life, in order to realize its dreams.[14] Dreams condition noetic life, as well as the dignity of a non-inhuman life in general.[15] As such they enable negentropic forms of knowledge and collective existence. In the face of the anthropogenic devastation of nature and the dead ends of political and social life, the need of a common noetic dream becomes pressing.

This dream necessitates intermittence. In order to dream, one must sleep, one must stop working. Intermittence conditions bifurcations: resisting incessant algorithmic automation, it suspends the already individuated individual, showing new modes of being and thus enabling transindividuation – a passage to a psychic and collective belonging and a passage to care and act: a transvaluation. In this analysis of the contemporary predicament, Nony's essay interweaves with great care the transvaluation effected by Stiegler's final act, his release through suicide, a release from the physical malady that afflicted him, as much as from the socio-technical malaise weighing ever-heavier. This transvaluation, Nony sums up evocatively in the figure of the flying fish, the symbol of *Pharmakon.fr*, the school Stiegler founded, an allegory of the noetic soul, intermittently below and above water, exploring and transforming milieus, opening new possibilities of life.

Gerald Moore goes further with the theme of intermittence, offering a thematic exposition of its philosophical, anthropological and historic provenance and its relevance for our socio-economic predicament. This predicament consists in the cancellation of intermittence by computational capitalism, in a twofold manner: (1) through a total surrender of productive time, driving the precariat to complete exhaustion under the threat of technological redundancy; and (2) through the inculcation of incessant, compulsive consumption on the basis of an algorithmic desire which demands, in turn, a total surrender of noetic space and lived time.

Moore rehearses Stiegler's appropriation of the Aristotelian theory of the soul and its tripartite appropriation into vegetative, sensitive and noetic layers.[16] For Aristotle, only the divine can lay claim to the incessant actuality of *noēsis*; all other beings must *live* intermittently. Thus, it is for Stiegler, that we enter a non-inhuman space only intermittently. However, beginning with the agricultural revolution, a splitting of population was set into motion, with the aim of creating segments that could operate without interruption either as material, or as knowledge producers and administrators. Whether then in privilege or subservience, the totality of the population was forced to relinquish rest and thought, both of which necessitate intermittence.

Computational capitalism and the automatization of society at large form thus the ultimate stage in a long process the last episode of which, as Moore shows, is Covid-19: a 'syndemic' determined by socio-economic structural patterns, as much as by biology. Against it, resilience as agility and adaptability will not suffice. New forms of desire and a transformation of work, that is, of technical life capable of building a future, are called for, in order to counter the impoverishment of biological, social and noetic diversity and establish complex negentropic loci.

The final essay of the section resumes and advances further the problematic of intermittence. Paul Willemarck begins his analysis from the tension between the uninterrupted mnemotechnic survival of Bernard Stiegler (books, films, documents, etc.) and the disappearance of the *idios* that he was. In order to make sense of this disappearance, of the ultimate failure of tertiary retentions to preserve life, Willemarck turns to the history of the project of *Technics and Time*. He departs not from the technical empowerment and extension of the human that this sprawling, unfinished work is often understood as advancing, but from the notion of *idios*, the quasi-human singularity, in its failure to exteriorize itself in tertiary retentions, to become text. This default of the *idiotext*, Willemarck identifies with Stiegler's originary default of the origin.

Within the Aristotelean division of the soul, the default of the *idiotext* corresponds to the noetic soul, embedded in the soul's vegetative and sensitive strata, and specifically to its constitutional failure to achieve an incessant mobility, a non-intermittent assimilation of its extended milieu. Showing that the vegetative and sensitive layers constitute the noetic soul as much as they are formed by it,

Willemarck shows that the default of the *idiotext* corresponds to the default of life *tout court*. Life, in its effort to take hold of death and extend itself via *noēsis* into tertiary retentions, finds itself in their grip, insofar as such retentions are ultimately 'death [seizing] hold of life'.[17] As such, it heals, but also perpetuates its originary, singular wound, which is no less a material wound of life, the ultimate figure of which is death.

Beyond entropy: Différance *and epochality*

The countering of death by organized life is premised on the possibility of countering entropy, or rather, on the possibility of deferring and utilizing its potential, for the sake of the formation of negentropic loci and functions. Stiegler's work, attentive to the need and exigences of fashioning such formations, passes from a *technological* to an *organological-pharmacological* and finally a *neganthropological* phase.[18] This passage entails significant implications for Stiegler's thinking on *différance* and epochality, which the essays of this section explore.

Stiegler's indictment that the efforts to overcome metaphysics have systematically disregarded the challenge of entropy and 'the pre-eminence of the thermodynamic question'[19] might only have come in the neganthropological phase of his late work, but an acute awareness of the pernicious ramifications of entropy accompanied its earliest pages. Georgios Tsagdis traces the development of this awareness through the three volumes of *Technics and Time* leading up to the outlines of the neganthropological project, aimed at countering Claude Levi-Strauss's despairing configuration of anthropology as 'entropology'.[20] This counter-move Tsagdis construes as a thinking of the time that remains.

In its remainder, his essay explores two questions or potentialities, which nuance negentropy in order to carry forth the neganthropological project. The first consists in recasting the equation of negentropy with 'a *différance*', that is, with 'a temporal deferral and a spatial differentiation of entropy'.[21] It proposes instead a formulation in which *différance* designates the *mutual* negentropic becoming of entropy and the entropic becoming of negentropy. The second potentiality consists in a series of 'translations', or as Stiegler prefers, 'bifurcations', which enable the play of *différance* to pass from the inorganic to the organic, to the life of *logos* and the algorithm. The potentiality of such negentropic translations can afford the ground for a series of distinct scientific and philosophical projects which remain to be undertaken in fidelity to Stiegler as a thinker of life.

The section proceeds with a sustained investigation of the constitution of the epochal and the role of entropy in determining its conditions of possibility. Stiegler frames Maurice Blanchot's insight that every change of epoch forces a turn of thought which cannot be accompanied by knowledge, as a technological and thus organological and neganthropological problem. Accordingly, our current epoch of computational disindividuation seems to make this temporary incapacity of knowledge permanent and with it to destabilize the 'we' that writes this epoch. Eric Hörl offers a compelling account of Stiegler's passage from a phenomenological, or subjective, to an 'objective' *epochē*,[22] driven by what Alfred Lotka termed exosomatization, the technical outsourcing of human functions, including ultimately that of thought. He is thus able to account for the disruptive predicament of the epoch Stiegler termed the Entropocene as 'an incapacity to achieve epochal redoubling'.[23]

If every epoch is both a transient phase of time and the suspension of this transition, just like every moment is both an instance of time, as well as its non-durational interruption, this is for Stiegler because technics effects always a suspension (*epochē*) of an epoch and thus ushers forth a new time.[24] This technological interruption is however not disruptive as long as knowledge can recuperate it, live up to it, enfold its implications and create new and diverse negentropic forms of thought and life; that is, as long as thought can achieve an *epochal redoubling*.[25] Pursuing Stiegler's diagnosis, Hörl demonstrates that the usurpation of thought by computational tertiary retentions and protentions, precludes such a redoubling, signalling thus not the passage to a new epoch, but the eclipse of epochality altogether. This is the 'non-epoch' of the Entropocene, 'an age of *indifférance*'.[26] Hörl's antidote, just like for Nony and Moore, consists on foregrounding the need for a 'care-ful' thinking, a thinking [*penser*] that heals [*panser*] and that is thus able to escape the 'melancholiform' imposed by the disruption of epochality.

The section concludes with an appraisal of what conditions and animates Stiegler's diagnosis of our current disruption of epochality. Donovan Stewart begins by rehearsing the critique Stiegler levelled against Heidegger and the history of metaphysics, as a repression of the originary technicity of the human. For Stiegler, the human emerges though technical *différance*, the 'transductive' co-constitution of the organic and inorganic, the *who* and *what*, life and technics.[27] This originary, constituting interplay gives time, it gives the epochal, creating calculable differences, and incalculable futures. As such, technical *différance* precludes the forces of epochal closure.

However, as Stewart demonstrates, Stiegler finds himself precariously close to the critique he levels against Heidegger, who was no less able perhaps to appreciate the foundational character of inauthenticity, as well as of calculability and clock-time.[28] Stiegler, in his quasi-apocalyptic diagnosis of the 'final stage of the Anthropocene' as 'accomplished nihilism', that is, as the 'absence of epoch',[29] comes very close to casting technics as a totalizing force, akin to Heideggerian *Gestell*. As Heidegger shows *Gestell* to strip beings of their differences, their *Geschlecht* (sex, gender, ethnicity, generation and nation at once),[30] Stiegler decries the entropogenic loss of biodiversity and noodiveristy. Stewart shows that this antagonism of an always originary and a now epoch-ending technicity forms an unresolved tension in Stiegler's thought from the start. Explicating the self-deconstructing potential of technical *différance*, he concludes that this tension does not constitute a misstep of Stiegler, but rather, a *hyper-ethical* decision.

Creative organologies: Works of invention

In an interview from 2015, Bernard Stiegler refers to organology as a 'both theoretical and practical' proposition.[31]

As a theoretical proposition, organology is general in that it concerns all theories. Yet, as a discourse on the conditions of possibility of all theories, it is at the same time an exposition on their conditions of impossibility: it is simultaneously an argument on the limits of all theories, and hence a statement on the necessity of practice.

This brings us to the definition of organology as a 'theory and practice' that is to be considered primarily as 'an *approach*': a way 'not only of posing questions, but of *letting oneself be put into question* (by technics, organs and organizations, combined and forming transductive relations) in order to confront [...] problems'.[32]

The papers assembled in this section explore the *practical* organological *propositions* Stiegler has set out, mostly in collaboration with others in organizations. These proposals present organological configurations that are supportive of processes of co- and trans-individuation. If we refer to these designs as creative, it is in order to emphasize how practical organology is not a matter of fabrication – of an object by a subject remaining external to the product – but about the strange interplay between the *who* and the *what* to which Stiegler refers specifically as *invention*.[33]

By way of introduction, Bart Buseyne explores the question of the philosophical import of Stiegler's commitment to the reinvention of the techno-cultural milieu and his concomitant sociopolitical involvement. As he discovers the milieu of spirit while in prison, Stiegler uncovers how *anamnēsis* does not perceive its technicity, in the same way that a fish cannot see the water, even though it is its element. This confrontation with *anamnēsis*' oblivion of its milieu urges him to elaborate the prime philosophical question of *anamnēsis* as the question of technics. In thus leading philosophy back to what he finds as its root, and establishing how we need to know what goes on in *hypomnēsis* if we are to know how to allow for thinking to unfold, his *passage to the* philosophical *act* passes into a concern with issues pertaining to the milieu. If Stiegler did go into so many elementary issues, he did indeed do so as a philosopher who explores questions to their (deconstructive) root.

In 2006 Stiegler established the Institute for Research and Innovation. He did so in close collaboration with Vincent Puig. With reference to projects he worked on with Stiegler, Puig demonstrates how Stiegler truly revolutionized the assessment of technological development in terms of its deproletarianizing potential. As the experimental sessions in computer-assisted improvisation organized at the 2015 cultural festival in Mons bring out nicely, Stiegler took a view on organological design that centred on the notion of the improbable 'to come' (*à venir*). While he approached technical instruments as objects of desire, to assess them from a pharmacological perspective, he would be very positively critical of what he thematized as the *incompletion* of *technologies of spirit* that do accommodate for the intermittence that should occur between the synchronic and the diachronic, the calculable and the incalculable. Puig concludes by outlining Stiegler's call to implement the requisite intermittence in the very economy of today's 'automatic society', and to set up a contributory economy between labour and work, *negotium* and *otium*.

The *contributive learning territory* project has been running since the autumn of 2016 in the Plaine Commune agglomeration north of Paris. As Maël Montévil explains, the programme emerged from the theoretical work of Stiegler and Ars Industrialis. The overall objective is to transform the economy into a contributive economy that is designed as a strategy to 'disrupt' technological disruption, and to initiate a comprehensive collective bifurcation through the development of knowledge in all its forms. Over the past years, several work groups have been established in the suburban area, while the set-up of some others is impending. The work that has been carried out so far focuses mainly on the economy, digital urbanism and young children's development in the context of the overuse of digital media. Montévil elaborates specifically on the latter, as he provides a perspective on the skills that the various participants have developed in interconnecting the work of academics, professionals and inhabitants.

The Internation Collective is an international group of researchers, academics, artists and citizens that was formed on 22 September 2018 at the Serpentine Galleries in London. The collective makes sure to submit proposals to the United Nations in order to rethink and reorganize work on new bases. Key to its commitment are two Stieglerian notions that Anne Alombert sets to explain in her paper: the internation and interscience. She first explores the political stakes of the internation as a project aiming at the creation of a network of localities that share the concern to design new economic models, so as to counter the toxic impact of the ways we are currently interacting with digital technologies. Alombert then surveys the epistemological stakes of the interscience as a project that works in particular towards the articulation of academic and extra-academic knowledges, so that universities and civil society can join forces in contending against the digital proletarianization of inhabitants. To end, she clarifies how both the contributory research programme running in Plaine Commune and the work done by the Internation Collective give concrete effect to the internation and interscience.

The Association of the Friends of the Thunberg Generation was created in the final days of 2019, at the initiative of Stiegler and the 2008 Nobel Prize laureate in Literature Jean-Marie Le Clézio, in close cooperation with internationally recognized scientists and several French environmental youth movements. As Victor Chaix spells out, in establishing the Association Stiegler wanted to insure the transfer of the knowledge, concepts, theories and visions developed and refined within Ars Industrialis and the Internation Collective, to younger generations, who are increasingly worried about the future, so that they may forge the noetic tools and armoury it takes to tackle the entangled issues of ecosystem destruction, technological disruption and sociopolitical fatalism. As he reflects on the Association's purpose, Chaix reads a number of Stiegler's notes on knowledge, intergenerational relations, and political action, to synthesize some of the most significant experiences and experimentations of the Association's young members as a *scholē*.

Between 2011 and 2014 Stiegler and colleagues at the Institute for Research and Innovation, the World Wide Web Consortium and the Unlike Us network set out to analyse social media platforms so as to build alternatives. Harry Halpin and Geert Lovink return to the work done at the time, to highlight this important, if somewhat underexposed episode in Stiegler's work. Tracing Facebook's lineage to Peter Thiel's reading of René Girard's concept of *mimesis*, they recall how its toxic effects on the mental condition of young users became clear fairly early. Determined to counter these, Stiegler and colleagues turned to Simondon to reconceive of 'users' as potentialities in a process of collective individuation. By removing the individual as a mere 'node' in the 'social' network, they sought to reattach the social to an information architecture that could further the formation of genuine processes of co- and transindividuation. To make this approach concrete, and give proof that another social world is possible indeed, Halpin and Lovink discuss a politics of knowledge that organizes specifically toward long-term thinking, called *noopolitics*.

Echoes: Individuating art

Stiegler was *un passioné des arts* who played his part to restore the passion in the arts.[34] He believed in the arts, and even considered that works of art only work 'unless one believes in them', since they 'work by indicating a plane other than existence', the plane of the improbable and indemonstrable that may 'open up' the public 'to address and destine it'.[35]

In a paper published in 2008, Stiegler restated Immanuel Kant's 'question of reflective judgement' in terms 'of the singularity of desire and the desire for the singular', to add that these 'can only be constituted in a relationship with technique'.[36] This recasting led him to the somewhat remarkable definition of 'technique' as 'the unhoped-for coincidence between technique and singularity'.[37]

That the singular individuates in a concurrence with technique does not take away from the fact that technique, like facticity, is at the same time 'what undoes all singularity'.[38] Since individuation is inhabited[39] from the start by 'the impersonality of a pre-individuality',[40] it is marked structurally by the default of technique, and can hence only occur in an unachievable negotiation with what erases it.

Unable to start from a pre-given originarity, individuation can only believe in its 'default as its origin'.[41] The fact itself that it can only occur in response to its defect and wound[42] brings us to the figure of Echo, the nymph whose voice lies in the death of the living voice. Is her tragedy not the tragedy of all speech, all literature, art, address and destination?

Echoes reverberate when it is too late to utter a word of origin. As she commits to the unhoped-for, Echo breathes 'an excess of signification',[43] to answer with nothing short of a declaration of love. The two essays that follow explore lovingly how the exceeding powers of repetition and recording may yield passionate anamnesic performances, as the singular comes to individuate in a differantial conjunction with technique, through felicitous, if finite and not finally ascertainable exertions of its default that echo as love: 'loving memory', 'love of music'.

In 2014, Stiegler lent his voice in an essay-film by Mischa Twitchin. Reading aloud from published work, he discussed an organology of memory, in a juxtaposition with images of gravestones rendered as books, evoking 'loving memory'. In his contribution, the director returns to Stiegler's voice (as it now returns to him), allowing a question of thinking, perhaps, to return to itself. For Stiegler, beyond being simply 'an academic discipline', philosophy 'is firstly a way of living'.[44] Reflecting on this as a question of and for care, Twitchin considers the voice of philosophy through a discrete instance of phonography. He examines the relation between care and technics – as, precisely, that of mnemo-technics – through a reading of *anamnēsis* and *hypomnēsis*, as it returns through the thought of 'loving memory' when relating the living and the dead.

Susanna Lindberg presents the 'heuristic privilege' that Stiegler accords to music. If Stiegler takes a particular interest in music among the arts, she argues, that is because it is a perfect illustration of a core idea in his early work, as developed in *Technics and Time*: as a marked instrumental art, music is time made sensible. Since it can touch sensitivity directly, without the mediation of *logos*, music is even a prime example of the affective effects of technics. Lindberg first discusses Stiegler's reinterpretation of Husserl's description of internal time consciousness through the example of a melody, which Stiegler stretches to include melody recordings on different supports. She then explains how Stiegler considers the role music plays in community building, as it formats the affective basis of the soul. Next, she comments on his analyses of the way in which today's culture industries configure and undo people's sensibility. Music illustrates very well how an industrial aesthetics comes to control the rhythms of bodies and souls. To end, she presents the emancipatory potential that is inherent to contemporary music technologies, according to Stiegler, while both professional and amateur musicians learn to play with these, to practice a truly giving 'love of music'.

An unfinished conversation

The closing section includes a single essay: the discussion Jean-Hugues Barthélémy and Bernard Stiegler have been engaged in, ever since they first met and befriended in the early 1990s.[45] Both authors carried philosophical projects that were very close to one another, and not only because they were both admirers and connoisseurs of Gilbert Simondon or André Leroy-Gourhan, but because they shared a passionate interest in the question concerning the non-orginarity of the ego.[46] Their intense collaboration stretches over more than two decades and has written an important chapter in the recent history of philosophy in France. One aspect of their ongoing argument relates to what Barthélémy calls the architectonic of philosophy, which is concerned with the foundation of law. This undertaking requires a reflection on the secondariness of the thinking subject, an archi-reflexivity that initiates a break with the anthropogenetic conception of philosophy. Whereas Stiegler considers language to be part of technics, Barthélémy advocates the conception of a philosophical relativity that takes account of the multidimensional character of sense in terms of the interpenetration of language and technics. This argument results in a discussion as to the sense which has to be given to the contributions of Simondon and Leroi-Gourhan.

Notes

1. Bernard Stiegler, 'To Love, to Love Me, to Love Us: From September 11 to April 21', in Bernard Stiegler, *Acting Out*, trans. David Barison, Daniel Ross, and Patrick Crogan, Stanford: Stanford University Press, 2009, 37–82, 39.
2. Stiegler, 'To Love, to Love Me, to Love Us', 56.
3. Stiegler, 'To Love, to Love Me, to Love Us', 59.
4. Bernard Stiegler, 'What Is Called Caring? Thinking beyond the Anthropocene', in Bernard Stiegler, *The Neganthropocene*, ed., trans., and with an introduction by Daniel Ross, London: Open Humanities Press, 2018, 188–270, 233.
5. Bernard Stiegler, 'How I Became a Philosopher', in Stiegler, *Acting Out*, 1–35; the text is based on a lecture Stiegler gave in the Spring of 2003, at the invitation of Marianne Alphant.
6. The address took the form of a 'Letter to António Guterres' and was signed by Bernard Stiegler and Hans Ulrich Obrist on the 11th of November 2019 in view of the Internation/Geneva 2020 Collective. It is reprinted in Bernard Stiegler & The Internation Collective (ed.), *Bifurcate: 'There Is No Alternative'*, preceded by a letter from Jean-Marie Le Clézio, with an afterword by Alain Supiot, and a lexicon by Anne Alombert and Michal Krzykawski, trans. Daniel Ross, London: Open Humanities Press, 2021, available at: http://www.openhumanitiespress.org/books/titles/bifurcate/.
7. See also Vincent Puig's essay in this volume: 'Taking Care of Digital Technologies'.
8. Bernard Stiegler, *Technics and Time, 1: The Fault of Epimetheus*, trans. Richard Beardsworth and George Collins, Stanford: Stanford University Press, 1998, 1.
9. Stiegler, *Technics and Time, 1*, 18.
10. Stiegler begins to engage with the threat of the transhumanist ideology, around the year 2016. This volume does not include a thematic analysis of his criticism of the movement, although references to it are made in the texts of Jean-Luc Nancy and Victor Chaix. Stiegler himself discussed transhumanism in *The Age of Disruption* in relation to the problem of hubris. The forthcoming fourth volume of *Technics and Time* is expected to delve further into the relation of posthumanism to the post-truth era and should offer material for further investigation.
11. Bernard Stiegler, 'Escaping the Anthropocene', in Stiegler, *The Neganthropocene*, 51–61, 54; Bernard Stiegler, 'Five Theses after Schmitt and Bratton', in Stiegler, *The Neganthropocene*, 129–38, 127.
12. Stiegler, *Technics and Time, 1*, 136.

13 Patrick Crogan & Bernard Stiegler, 'Knowledge, Care and Trans-Individuation: An Interview with Bernard Stiegler', *Cultural Politics* 6/2 (2010): 157–70.
14 Bernard Stiegler, 'Elements of Neganthropology', in Stiegler, *The Neganthropocene*, 76–91, 76.
15 Bernard Stiegler, *The Age of Disruption: Technology and Madness in Computational Capitalism*, followed by *A conversation about Christianity with Alain Jugnon, Jean-Luc Nancy and Bernard Stiegler*, trans. Daniel Ross, Cambridge: Polity, 2019, 18.
16 Bernard Stiegler, *The Decadence of Industrial Democracies. Discredit and Disbelief, Volume 1*, trans. Daniel Ross & Suzanne Arnold, Cambridge: Polity, 2011, 133–6.
17 Stiegler, *The Decadence of Industrial Democracies*, 147.
18 Daniel Ross, 'Introduction', in Stiegler, *The Neganthropocene*, 7–32, 22.
19 Stiegler, 'What Is Called Caring? Thinking beyond the Anthropocene', 202.
20 Claude Lévi-Strauss, *Tristes Tropiques*, trans. John Russell, New York: Criterion Books, 397.
21 Bernard Stiegler, *The Nanjing Lectures 2016–2019*, trans. Daniel Ross, London: Open Humanities Press, 2020, 303.
22 Bernard Stiegler, *Technics and Time, 2: Disorientation*, trans. Stephen Barker, Stanford: Stanford University Press, 2009, 7.
23 Stiegler, *Technics and Time, 2*, 7.
24 Bernard Stiegler, 'Programs of the Improbable, Short Circuits of the Unheard-of' (1986), *Diacritics* 42/1 (2014): 70–109, 84–5.
25 Stiegler, *Technics and Time, 2*, 60.
26 Stiegler, 'What Is Called Caring? Thinking beyond the Anthropocene', 268.
27 Stiegler, *Technics and Time, 1*, 177.
28 Martin Heidegger, *Being and Time*, trans. John Macquarrie & Edward Robinson, New York: Harper and Row Publishers, 2008, 373.
29 Stiegler, *The Age of Disruption*, 7, 12; Stiegler, 'Elements of Neganthropology', 81; Stiegler, 'Escaping the Anthropocene', 53.
30 Martin Heidegger, *Bremen and Freiburg Lectures: Insight into That Which Is and Basic Principles of Thinking*, trans. Andrew J. Mitchell, Bloomington: Indiana University Press, 2012, 34.
31 Judith Wambacq & Bart Buseyne, '"We Have to Become the Quasi-cause of Nothing – of Nihil": An Interview with Bernard Stiegler', trans. Daniel Ross, *Theory, Culture & Society* 35/2 (2018): 137–56, 141.
32 Wambacq & Buseyne, '"We Have to Become the Quasi-cause of Nothing – of Nihil"', 141.
33 Stiegler, *Technics and Time, 1*, 137.
34 For the notion of *passion*, see Bernard Stiegler, *Philosophising by Accident: Interviews with Élie During*, ed. and trans. Benoît Dillet, Edinburgh: Edinburgh University Press, 2017, 38.
35 See the short extract from a lecture Stiegler gave at an unspecified occasion, entitled 'Mystagogie et art contemporain', dated 25 April 2013, and published on the website of Cie Hendrick Van Der Zee, available at: http://www.hvdz.org/blog/archives/10760/art-contemporain-un-tres-court-extrait-de-b-stiegler-sur-mystagogie-lart-contemporain. Let us mention that Stiegler's aesthetics is not – contrary to what the above paraphrase may suggest – a 'reception aesthetics' in distinction to a 'creation aesthetics': reception calls for creation in that it must find expression, while creation is receptive in that it does not start from a pre-given origin.

36 Bernard Stiegler, 'Mystagogy: On Contemporary Art', in *Thinking Worlds: The Moscow Conference on Philosophy, Politics, and Art*, ed. Joseph Backstein, Daniel Birnbaum, Sven-Olov Wallenstein, London: Sternberg Press, 2008, 31–46, 38.

37 Stiegler, 'Mystagogy: On Contemporary Art', 39. The definition is reminiscent, in its illogicality, of Vladimir Nabokov's description of beauty as '[b]eauty plus pity' (Vladimir Nabokov, *Lectures on Literature*, ed. Fredson Bowers, Introduction John Updike, London: Pan Books, 1983, 251). Stiegler's characterisation of technique as *technique* plus *singularity* is consistent with his conception of textuality that states that a text only occurs in an individuation: *textuality* is *textuality* plus *idios*, or in other words, *idiotext* (cf. Bernard Stiegler, 'What Is Missing/What Makes Faults' (Draft only), trans. Daniel Ross [40 p.], 13, available at: https://www.academia.edu/54731473/Bernard_Stiegler_What_is_Missing_What_Makes_Faults_Ce_qui_fait_d%C3%A9faut_1995_).

38 Stiegler, 'Mystagogy: On Contemporary Art', 39.

39 Bernard Stiegler, 'Wanting to Believe: In the Hands of the Intellect', in Stiegler, *The Decadence of Industrial Democracies*, 131–62, 152, 161.

40 Stiegler, 'Wanting to Believe', 161.

41 Stiegler, 'Wanting to Believe', 161.

42 Stiegler, 'Wanting to Believe', 158.

43 John Sallis, *Echoes: After Heidegger*, Bloomington and Indianapolis: Indiana University Press, 1990, 1.

44 Stiegler, *The Age of Disruption*, 75.

45 Bernard Stiegler, *Technics and Time, 3: Cinematic Time and the Question of Malaise*, trans. Stephen Barker, Stanford: Stanford University Press, 2011, 220.

46 Stiegler, *Technics and Time, 1*, 252.

Part I

IN MEMORIAM: BERNARD STIEGLER

Chapter 2

NOBLE NEGANTHROPOLOGIST: REMEMBERING
BERNARD STIEGLER (1 APRIL 1952–5 AUGUST 2020)

Pieter Lemmens

When a friend mailed me on the evening of the 6th of August this year that he got informed via Facebook that Bernard Stiegler had died, I was first unwilling to believe it. As a deliberate non-user of Facebook or any other electronic grapevine that goes for 'social media' nowadays, I immediately went online to check for confirmation on some website but could not find any yet. Two hours later, though, I received another e-mail from a British colleague closely related to Stiegler that conveyed the same sad message. I was flabbergasted, completely at a loss. How could it be that this great philosopher, this exceptionally brilliant and productive thinker, this incredibly prolific author and tireless intellectual activist who taught and inspired so many thinkers young and old all over the world, and who was truly at the height of his career and still full of ambitions, had suddenly passed away, at age sixty-eight, way too early? All the projects that he was still involved in, all the books and further volumes that were still to be written or completed, all the promising horizons that were radiating from his texts and from his amiable and admirable person? He was the originating, inexhaustible driving force of so many things! His voice, his sharp intellect, his warmth, his humility, but also his severe prophecy, his shining example of a genuine philosophical *vita contemplativa et activa*. All this will be tremendously missed by everyone who knew and cooperated with him, and his passing represents a great loss for the philosophical community.

With the death of Stiegler, a genuine *maître penseur* of our time, philosophy has lost one of its true giants. I still feel the eerie silence that the sudden arrest of his vibrant and exceptional *furor philosophicus* has left behind – right in the gloomy midst of this ongoing global pandemic, isolated at home 'in a "post-truth" world where no one trusts anyone anymore', as he wrote in a memo on Covid-19 on the 10th of April 2020. And with each passing day in this desolate and desperate time of post-truth, fake news and conspiracy paranoia running rampant, Stiegler's analyses of the hyper-toxicity of the contemporary technical milieu of the mind and in particular his ominous warning about the symbolic order becoming diabolic, ring true ever more frighteningly.

I was never very close to Stiegler, but our few encounters have left me with some precious memories. Before meeting him in person for the first time more than ten years ago, I was struck by his work, which blew me away but also enabled me for the first time to truly understand Heidegger, that infinitely fascinating dark sphinx from the Black Forest. I first encountered his work in late 2005, when I more or less *accidentally*[1] stumbled upon the English translation of the first volume of *La technique et le temps*[2] in the fine library of our philosophy faculty at Radboud University in Nijmegen, which is nowadays mostly deserted. I was at the time writing my dissertation on the implications of the biotechnology revolution for the human condition. As someone initially trained as an evolutionary biologist with a special interest in human evolution but having become hopelessly infected with Heideggerian phenomenology and existential ontology during my philosophy study, I was particularly puzzled by the question of how something like *Dasein* or the human as an existing being could have ever come into being through a biological process of evolution.

At the time I had already learned from Peter Sloterdijk's notorious 1999 lecture 'Rules for the Human Park' and later from the chapter 'The Domestication of Being', both to be found in his 2001 book on Heidegger, *Not Saved: Essays after Heidegger*,[3] that the human, in terms of what Heidegger called *Dasein*, could and should be thought of as being the result *not* of a process of biological evolution but of technological evolution, that is, of *technogenesis*, a term that was also used by Stiegler as I later found out while reading *Technics and Time, 1*. It was only after reading the book several times that I started to grasp its profound meaning and the enormity of the project that it aimed to initiate. I began to appreciate *Technics and Time* as the landmark treatise that it is on a par with books such as *The Phenomenology of Spirit*, *Logical Investigations* and of course *Being and Time* itself, to which it explicitly responds. Alas, the *Technics and Time* project will now remain *unvollendet* forever, although at least parts of it may still appear in some form, since Stiegler was constantly in the process of completing them.[4]

Having read, almost in exuberant intoxication, all of Stiegler's work that was available, I watched in 2007 the mesmerizingly beautiful documentary *The Ister*[5] by the Australian philosophers Daniel Ross and David Barison, which had Heidegger's 1942 lecture course on Hölderlin's eponymous hymn as subject, but in which a still rather young and remarkably *flamboyant* Stiegler featured most prominently, flanked by his older compatriots Jean-Luc Nancy and Philippe Lacoue-Labarthe as well as the German cinematic genius and *enfant terrible* Hans-Jürgen Syberberg.

I first met Stiegler in the flesh on the 14th of September in 2010, when I visited him for an interview at his office at the *Institut de Recherche et d'Innovation* (IRI) in the nondescript white apartment block opposite of the colourful Centre Pompidou in the bustling centre of Paris. I was very happy at the time that he had allowed me to interview him and that he was also willing to do that in English, as I was aware that French would be much more convenient for him. As I remember, he had his residence at the very top floor of the building, on a kind of loft where he sat at a huge desk overseeing the IRI-employees working below. What struck me immediately when we started to talk was his great friendliness and warmth, his low-key approachability and the complete lack of arrogance or 'attitude' that so many lesser minds in the academic world so frequently exhibit.

I had come in particular to question him about his thoughts on open source and commons-based practices of innovation and production in the digital- and bio-economy, the subject of my post-doc project at the time, from his perspective of the 'economy of contribution'. Soon enough though we were engaged in a discussion on the more general aspects of his views on technology as well as on the current tendency of capitalist-controlled digital innovation to produce increasing stupidity, irrationality, proletarianization, addiction and what he called symbolic misery. When talking about addiction he confessed that he had been an addict himself once and he referred to *Naked Lunch*[6] by William Burroughs, a novel about heroin addiction which, he argued, conveyed like nothing else the true nature of addiction, and that is: the impossibility to stop (using) despite desperately hating it and wanting to stop (using). 'Capitalism is now confronted with this problem', he argued. The truth of this insight has become ever more apparent since then of course. Only Covid-19 has managed to cause at least a temporary and partial halt to the ruthless and rabid capitalist juggernaut that is tearing the world apart and forced people to some modicum of abstinence, but there is no sign yet of anything even remotely resembling the willingness to fundamentally change habits, let alone of what Heidegger called *Besinnung*.

Obviously, Stiegler first of all had consumerist addiction in mind, probably not knowing already at the time, as neither I did, about the extent of state-sponsored opioid addiction in the United States, which has been escalating in the last decade. He did relate, though, that over 15 per cent of the American youth regularly used drugs such as Ritalin and Prozac, and without doubt this has also been on the rise, the obvious reason being, as he asserted, that 'this system does not produce pleasure anymore'.[7] Stiegler alluded to Burroughs in his books only on a few occasions as I recall, but I have the feeling that the writings of this arch-junkie and greatest intellect of the beats had a significant influence on his philosophical engagement. At least I have always sensed a more than coincidental affinity with Burroughs's views on language as a virus and his scathing critique of capitalist society in general and American capitalism – or 'the American nightmare' – in particular.

I began corresponding more often with Stiegler in 2013 when I translated his volume of interviews with Élie During, *Philosopher par accident*[8] in Dutch. When the book came out, I first invited him to the Netherlands for a lecture at our university, which he gave in a lecture hall fully packed with students, who apparently had a hard time following his train of thought, which largely flew over their heads due not only to its intrinsic complexity but also because of his now already legendary, in Nijmegen that is, 'Fringlish' (French-English). Since that first visit, I invited him three more times to Nijmegen for lectures or participation in seminars. In a gesture of reciprocation, I was invited in August 2014 to the *Pharmakon* summer school, which he organized every year since 2011 together with the Ars Industrialis association in Épineuil in the middle of France, on the estate of his home residence.

There are two things that I remember especially from these week-long events in the sunny French countryside, which were attended not only by professional philosophers from all over the world (many of them young and some camping on the grass near his house), but also by local citizens, something Stiegler very much encouraged. One is the enormous energy and spirit with which Stiegler animated the discussions as well as the extensive, thoughtful, indeed carefully crafted comments that he always offered to *each* speaker at the end of the day's lengthy sessions. The other is the friendly and immensely joyful atmosphere during the long evenings when everybody gathered around the large wooden tables in the garden around the house to share a most excellent dinner, drink delicious wines, engage in impassioned philosophical conversations sometimes until deep in the night and get to know each other more intimately. It was there that I first met Yuk Hui, Paul Willemarck, Gerald Moore, Anaïs Nony, Sara Baranzoni, Paolo Vignola and many other members of the Stiegler crowd. It sometimes felt as if I had landed in the garden of Epicurus.

In June 2016 I organized a meeting between Peter Sloterdijk and Stiegler in collaboration with Yuk Hui and the Nootechnics collective on the topic of the Anthropocene. Besides a two-day academic seminar with a choice of international speakers, we also arranged a big public dialogue in which Stiegler and Sloterdijk crossed swords on the Anthropocene. This theatrical event was held in the great concert hall in the city centre and completely sold out with over 1500 people attending. Despite its great success in terms of publicity and attendance, the 'harvest' of the event proved rather poor philosophically, as Sloterdijk largely refused to engage in a sincere dialogue, instead choosing to entertain the audience. This starkly contrasted with Stiegler's admittedly over-enthusiastic and also somewhat schoolmasterly attempts to teach the audience about the gravity of our predicament in the Anthropocene. Today, more than four years later – and after the famous, 'viral' speech by the fifteen-year-old Swedish schoolgirl Greta Thunberg (prominently compared to Antigone by Stiegler in his last book)[9] at the COP24 UN Climate Change Conference in December 2018 in Katowice, Poland, and after the first protests of Extinction Rebellion in the same year – the public awareness of that predicament has significantly increased, although we're still awaiting a decisive 'social tipping point'.

Something similar happened in the seminar. While Stiegler was very eager and willing to enter into a genuine debate with the Colossus of Karlsruhe, constantly probing and provoking him, the latter persisted in his typical, humorous yet distanced monologues and only at the very end managed to give some impression of being engaged in a dispute. Stiegler had also prepared notes for the event and had dealt with Sloterdijk's theory of 'disinhibition' from the latter's 2005 treatise *In the World Interior of Capital* in his then just published book *Dans la disruption: Comment ne pas devenir fou?*,[10] a copy of which we also sent to Sloterdijk, accompanied with a personal letter. There is no sign at all that Sloterdijk has ever made any effort to an answer. All in all, I think it was largely a failed encounter, due to Sloterdijk's unseemly haughtiness and recalcitrance. I still remember Stiegler telling me back in 2010 in Paris that he considered Sloterdijk to be a sophist, 'a brilliant sophist, very brilliant, but a sophist nevertheless', as he conveyed.

In January 2018, I invited Stiegler again to Nijmegen for an encounter with the doyen of American philosophy of technology, the post-phenomenologist Don Ihde, about the future of philosophy of technology in the age of the Anthropocene. Although coming from pretty similar backgrounds philosophically and having 'technologized' phenomenology, each in their own particular way, Ihde and Stiegler had quite different opinions on how to think and approach technology in the current conjuncture of global ecological crisis and ubiquitous digitalization, based on widely diverging diagnoses of our predicament. There was a lively and friendly debate though and thanks also to the presentations of all the other contributors it was a great success. Yet again, it was remarkable during this seminar to perceive the stark contrast between Stiegler's deeply engaged, combative style and *esprit de sérieux*, and the more relativizing, pragmatist, relaxed attitude in debating characteristic of Ihde, sprinkled with ironic and joking digressions. When Stiegler spoke though, with a solemn voice, the audience was particularly silent and got to listen carefully to an entirely new and utterly original interpretation of the later Heidegger's complex notions of *Gestell*, *Ereignis* and *Kehre*.

In my undoubtedly simplifying understanding, Stiegler in some sense did to Heidegger what Marx did to Hegel: turning his thought around as it were and putting it on its feet, that is, in his case by offering a materialist re-interpretation of Heidegger's still transcendentalist or even idealist understanding of human existence and of what the latter called the question of Being. What he showed in great detail, and that is to say through meticulous phenomenological meta-analyses of Husserl's transcendental phenomenology and especially of Heidegger's existential ontology, is that the temporal and historial dynamic of *Dasein*'s mode of being as being-in-the-world is fundamentally, and that is to say *originally*, constituted and conditioned by technology in the ontic sense. Existential time as understood by Heidegger is constituted by technics and the human condition as 'originally being in default' is therefore a technical condition through and through. This is Stiegler's great insight, which he has elaborated ever since, applying it not only deconstructively but also reconstructively to virtually all dimensions of human existence (from philosophy, politics, ethics, history and economics to science, education, anthropology, psychoanalysis and the arts) and taking it as the key to a deconstructive reading of the Western philosophical tradition as a whole from Plato to Heidegger and his own teacher Jacques Derrida, whose most original and critical reader he had no doubt become.

Like Sloterdijk, but in a much more profound and incisive manner, Stiegler combined Heideggerian existential ontology and the thinking of Being with (paleo) anthropology (in his case drawing on André Leroi-Gourhan) to show how the human's openness to Being and understanding of beings had emerged and evolved through a process of technical exteriorization (and biological interiorization) and how the history of truth in the sense of Heidegger can be understood as resulting from successive adoptions of the ongoing transformations of technical systems as they periodically – and today in the 'age of disruption' permanently – re-condition the human condition.

While the first volume of *Technics and Time* laid most of the theoretical groundwork for his project by demonstrating why our time and most decisively our future should be thought of as conditioned by the specifics of its technical prostheses, the second volume, which addressed the current 'disorientation', showed how this conditioning had taken place throughout the history of the West and why the current technical prosthesis, that is, digital technology in the context of global capitalism, obstructs proper adoption and instead produces adaptation. The third volume began to explicitly put this question into a politico-economic context and initiated the project of renewing the critique of political economy for the twenty-first century – as a critique of *libidinal* economy. It also started Stiegler's more active engagement with what he called technopolitics and noopolitics, which led in 2005 to the establishment of Ars Industrialis, an international association for the industrial politics of the mind.

In later works, from the 2003 essay 'To Love, to Love Me, to Love Us'[11] onwards, he always combined a philosophical diagnosis and prognosis with political and politico-economic proposals for a therapy, all the while elaborating, refining and expanding his massive and powerful conceptual arsenal (and we know he explicitly thought of concepts as weapons). This culminated eventually in what he called his organology and pharmacology of technology, which he began to theorize in terms of entropy and negentropy explicitly since the General Organology conference in 2014 in Canterbury, even if these notions have played a role in his work from the very beginning as any attentive reader will be able to confirm. Since 2016 he put his ideas on deproletarianization and contributory economics in practice through the *Plaine Commune* project in the Parisian suburb of Seine-Saint Denis, in cooperation with the municipality and French technology companies Dassault and Orange.

Stiegler always insisted that he was not a philosopher of technology in the usual sense of the word, that is, someone who thinks about technology as a specific object domain. For Stiegler technology, being the very condition of thinking and indeed of existing *as such*, was nothing less than *the* object of philosophy par excellence, that is, the original yet originally repressed (in the history of metaphysics) question of philosophy. He thus asked the question of philosophy in the same way that Heidegger asked the question of Being and he claimed that more original than the forgetting of Being was the forgetting of technology, since any question of Being by *Dasein* presupposes its prior *being put into question by* technology. Philosophy itself, he argued, originated in ancient Greece from what could be called a pharmacological 'trauma', that is, from the disruptive introduction of alphabetic writing technology in oral Greek culture and its ambiguous effects on the Greek psyche and society, soon to become the *polis*. Philosophy started off as an attempt to properly adopt the *pharmakon* of writing, that powerful magic potion interfering with the mythic mindset, as a medicine for the mind, and understood itself as a 'therapy for the soul'.

This is also why for Stiegler the first question of philosophy is not that of Being but of *teaching*, of teaching the *practice* of philosophizing as it is based from the outset on the *pharmakon* of writing, although precisely *this* has been consistently repressed from Plato onwards. A similar repression largely persists in today's institutions of learning with respect to the 'digital disruption', to which capitalism has responded much more profitably (although self-destructive in the longer run as we are now in the process of experiencing) – to the total detriment of the life of the mind. It is in this context that Stiegler called for a new 'battle of intelligence' to fight the current 'denoetization' and work to usher in a 'noetic renaissance' for the digital and ecological age, which is so desperately needed in our dire times of total nihilism, post-truth, systemic stupidity and generalized corruption of the financial and political spheres.

Stiegler was the eminent thinker of crisis and catastrophe and for him the time of crisis was *the* time of philosophy. Every great philosophy is at heart a response to a situation of profound crisis, confronting it at its height, be it that of Plato, Augustine, Descartes, Kant, Hegel, Marx, Nietzsche Husserl, Heidegger or Lyotard. And every crisis demands a new *critique* as well as a new *judgement* – *krinein* in Greek – based on new *criteria*. In my opinion no thinker was more aware of the sheer gravity of our current crisis and the need for an entirely new critique than Stiegler. His diagnoses may have often been bleak and dark yet they were also frighteningly accurate, as far as I'm concerned. No one probably understood better the magnitude of the challenge that we are confronted with in 'our' time, a time which 'we' have not yet properly managed to *make* 'ours'. *Who* that 'we' will be, will ultimately depend on *how* it will 'find' itself and respond to its new planetary condition technologically and societally.

The new name of this crisis of planetary proportions that is far more radical than the Axial Age has become the Anthropocene since the turn of the century. When Stiegler imported this notion in his thinking in 2014, he characterized it as the *Entropocene*, as the age of massive and generalized entropization of the biosphere, the most familiar symptoms of which are global warming, loss of biodiversity and ecological destruction. Yet he emphasized that this entropization first of all meant the loss of psychodiversity, sociodiversity and primarily noodiversity, arguing that the ecological crisis had its ultimate roots in a libidinal crisis, induced by capitalist control and exhaustion of the libidinal economy. Our response to it should consist in attempting with all our efforts a pharmacological metamorphosis of the global organological arrangement, which he began to call the technosphere in 2018. The goal should be to transform this technosphere again into an engine of terrestrial negentropization, that is, of the re-emergence and flourishing of diversity at all levels but first of all to re-ignite desire and forge a new *will to know* and *will to care* for the planet. He envisioned this as the advent of what he called the Neganthropocene, and with it, the birth of a new human being, the *neganthropos*, who is explicitly aware, and explicitly takes care, of its precarious thermodynamic inclusion – *as* techno-noosphere – in the Earth's biosphere.

As a heir of the enlightenment, Stiegler rejected Heidegger's 'mystagogic' hope for a *god* who could still save us from our persistent captivity in enframing, yet he was lucid and realistic enough to admit that the *bifurcation* from the Anthropocene into the Neganthropocene would entail no less than a *miracle* given the arguably desperate situation in which we find ourselves. Candidly conceding his depression about the state of our world, or rather our 'unworld' (*immonde*), he confessed in the *Age of Disruption*: 'I am often overwhelmed because it seems to be *absolutely irrational* to *believe* that a positive bifurcation could arise from out of the chaotic period in which we are rushing at the high speeds imposed by disruption. It is *totally improbable*. And this is a motive for despair.'[12]

Yet he summoned us precisely to actively believe in that improbability, and that is to say in a *miracle*. After the death of God and the unlikelihood of a return of the gods we should think of this miracle in terms of negentropy or rather neganthropy, of the glaring improbability that it represents in the universe, an improbability that nonetheless thrives on this planet, having originated and sustained life and allowing for the utterly improbable lifeform that is *Homo sapiens*, that dazzling 'festival of negentropy'[13] as Sloterdijk has it, having now become the prime generator of entropy on the planet though. If the coming of the *neganthropos* is the *improbable*, indeed maybe the *impossible*, then we should, with all our effort, *strive*, with and in memory of Stiegler – noble neganthropologist – for that impossibility and take at heart with Sloterdijk that 'the true realism of the species consists in not expecting less from its intelligence than what is demanded from it'.[14] For there is indeed, most probably, no alternative.

The ominous year of 2020 may well be called an *annus miserabilis* if not an *annus diabolis*. It may yet also prove to have been a turning point. Let us therefore actively believe, with the same fervour that animated Stiegler, that we may once welcome another *annus mirabilis* in an incalculable future that we can *still* imagine and dream of. And may we have the *courage* to do so, cultivating it among ourselves and following the shining example of Stiegler, that paragon of an intellectual warrior who embodied that courage like no other and who taught us that philosophy is first of all a *battle for* intelligence and *against* stupidity. Bernard, thank you so much for being such a towering example for all of us! I salute you.

Notes

1 An important notion for Stiegler, who became a philosopher 'by accident' as he explains in his partly autobiographical book *Passer à l'acte* from 2003 (Paris: Galilée), translated in English and included in *Acting Out* ('How I Became a Philosopher', in Stiegler, *Acting Out*, trans. David Barison, Daniel Ross, and Patrick Crogan, Stanford: Stanford University Press, 2009, 1–35).

2 Bernard Stiegler, *La technique et le temps 1: La faute d'Épiméthée*, Paris: Galilée, 1994; Bernard Stiegler, *Technics and time, 1: The Fault of Epimetheus*, trans. Richard Beardsworth & George Collins, Stanford: Stanford University Press, 1998.

3 Peter Sloterdijk, 'Rules for the Human Park', in Peter Sloterdijk, *Not Saved: Essays after Heidegger*, trans. Ian A. Moore & Christopher Turner, Cambridge-Malden: Polity,

2013, 193–216; Peter Sloterdijk, 'The Domestication of Being', in Sloterdijk, *Not Saved*, 89–148.

4   That is, the fourth and maybe also the fifth, sixth and even seventh volume, as announced in the preface to the 2018 republication of the first three volumes with Fayard (Bernard Stiegler, *La technique et le temps 1: La faute d'Épiméthée – 2: La désorientation – 3: Le temps du cinéma et la question du mal-être suivis de Le nouveau conflit des facultés et des fonctions dans l'Anthropocène*, Paris: Fayard, 2018).

5   Daniel Ross & David Barison, *The Ister* (DVD), Fitzroy: Black Box Sound and Image, 2005.

6   William Burroughs, *Naked Lunch*, Paris: Olympia Press, 1959.

7   Pieter Lemmens, '"This System Does Not Produce Pleasure Anymore": An Interview with Bernard Stiegler', *Krisis* 1 (2011): 33–42.

8   Bernard Stiegler, *Philosopher par Accident: Entretiens avec Élie During*, Paris: Galilée, 2004; Bernard Stiegler, *Per toeval filosoferen: In gesprek met Élie During*, vertaling Pieter Lemmens, Zoetermeer: Klement, 2014; Bernard Stiegler, *Philosophising by Accident: Interviews with Élie During*, trans. Benoît Dillet, Edinburgh: Edinburgh University Press, 2017.

9   Bernard Stiegler, *Qu'appelle-t-on panser? 2: La leçon de Greta Thunberg*, Paris: Les Liens qui Libèrent, 2020.

10 Bernard Stiegler, *Dans la disruption: Comment ne pas devenir fou?*, suivi d'un *Entretien sur le christianisme* [entre] Alain Jugnon, Jean-Luc Nancy et Bernard Stiegler, Paris: Les Liens qui Libèrent, 2016; Bernard Stiegler, *The Age of Disruption: Technology and Madness in Computational Capitalism*, followed by *A conversation about Christianity with Alain Jugnon, Jean-Luc Nancy and Bernard Stiegler*, trans. Daniel Ross, Cambridge: Polity, 2019.

11 Bernard Stiegler, 'To Love, to Love Me, to Love Us: From September 11 to April 21', in Stiegler, *Acting Out*, 37–82.

12 Stiegler, *The Age of Disruption*, 303.

13 Peter Sloterdijk, *Foams: Spheres, Volume III: Plural Spherology*, trans. Wieland Hoban, New York: Semiotext(e), 2016, 711.

14 Peter Sloterdijk, *Weltfremdheit*, Frankfurt am Main: Suhrkamp, 1993, 381.

Chapter 3

STIEGLER, MELANCHOLY, NEGATIVITY (FUNERAL SONG FOR BERNARD)

Jean-Luc Nancy

1

The death of Bernard Stiegler is part of his work. This in itself is not a rare exception. Not only are other deaths part of the work of the one who has died, but perhaps every death can well be considered part of the oeuvre that all life secretes. Nevertheless, in the case of Bernard Stiegler it seems particularly important to discern the precise character of this belonging and how it extends the work, perhaps by causing it to bifurcate in a singular way.

Man's relation to death – to death insofar as for each one of us it is *our* death and for all of us the death of each one – plays an essential role in Stiegler's work since it is in this relationship that, from the outset – from the beginning of *The Fault of Epimetheus* – he sees the decisive aspect of what he calls the 'invention of the human',[1] making of man himself a kind of artefact that would precede and call for all possible artefactions – everything that we name technics. This originary character is therefore itself nothing original: on the contrary, it indicates a (de)fault of origin [*défaut d'origine*]. The human does not relate to an origin – which would doubtless not even create a relation, but a continuity, or even an extension: it relates to a default of origin that is revealed in and as the feeling of death. The feeling, which is to say at once the perception, the experience and the knowledge that nourishes, as he says, 'concern' in the strongest sense of the word: care, worry [*souci, inquiétude*] – if not angst [*angoisse*] – in the anticipation of a destiny both singular and finite – singular because finite and finite because singular.

The text was drafted for the memorial symposium that occasioned the composition of the present volume. It followed on from the paper that was presented three months earlier at the New York University colloquium in honour of Stiegler, and that has since been published under the title 'Mélancolie', in: J.-L. Nancy (ed.), *Amitiés de Bernard Stiegler. Douze contributions réunies par Jean-Luc Nancy*, Paris: Éditions Galilée, 2021, 107–12. The texts form a diptych of sorts.

This feeling of death – this feeling-oneself-mortal – brings, with the invention of the human, the invention by him of the Immortals, these gods with whom is initiated the relationship of lack and substitution that makes of the human a technician.

This feeling is 'disastrous' [*funeste*], he writes in a surprising manner – since it means that he effects that of which he is the witness. '*Funeste*' refers, indeed, to what bears misfortune and ultimately death, as shown by the related words 'funeral' and 'funerary'. Everything happens as if this feeling brings death or leads to it, even though it seems rather to be its imprint, trace or stigma.

In reality, in a single inaugural phrase – 'Everything will thus have come with the feeling of death' – Stiegler makes this feeling the origin of the human, a de-originated origin, so to speak, the origin of what is experienced without origin and consequently also without end (destination) other than its own end (cessation).

This feeling has a name: it is melancholy. Stiegler writes, 'into the disastrous feeling of death, into melancholy'. We are only on page 141 of the book.[2] The motif of melancholy – I mean of this precise term – will not be examined any further, but later we read – after having passed through the 'primordial melancholy' symbolized by the interminable devouring of Prometheus' liver – that our 'always distanced proximity' to the Immortals initiates 'an infinite regret through which the eternal melancholy of the *genos anthropos* is woven'[3] – an expression as strong as that of the 'disastrous feeling'.

Melancholy will occasionally be discussed again in *Technics and Time* – for example, in a manner inspired by Barthes, in relation to the melancholy of the photograph. In his preface to the 2018 republication of *La technique et le temps*, Bernard, recalling work previously undertaken on Dostoyevsky's *The Idiot* and announcing that this motif will be that of the intended final volume (*Le défaut qu'il faut. Idiot, idiome, idiotie*), evokes epilepsy, 'one of those mental illnesses that can give birth to genius' – while adding, 'like melancholy according to Pseudo-Aristotle'[4], and he adds the reference to the publication of the text (which, incidentally, is recognized as indeed being by Aristotle). A detail, but one that is perhaps – or perhaps not – remarkable: he writes, in the title to which he refers, *mélancholie*, an old spelling[5] that the translator-commentator of the text, Jackie Pigeaud, had in the Latin form *Melancholia* used as the title of another work. As if Bernard was involuntarily indulging in a slightly archaicizing shift as far as the question of the relationship to the lost origin is concerned.

2

Here, it is not this trait that matters, but that of genius. Genius will return in the text – as the 'genius of the arts' (to translate *entekhnē sophia*) that Prometheus stole from the Immortals, and then as the 'genius of the Mesopotamians, older, *more originary* than that of the Greeks'. From one to the other, from myth to a protohistory revisited in order to push back the supposed 'Greek miracle' (and one can imagine putting it back indefinitely), genius here bears the mark (in accordance with its name, if one thinks about it, but Bernard does not make this explicit) of an anteriority in the origin that implies a trace of immortality in human genius. If we keep in mind the role played by the 'idiot' – a figure close to genius – idiocy is 'an originary disorder: an *originary exception*'. It represents in the default of origin a 'carelessness, this primordial idiocy, source of finite singularity and freedom'.[6]

There would thus be, alongside the default of origin or within it, even as the default itself, the resource of another initiality, that of invention and more generally that of acting (Bernard says), which I allow myself to gloss as the initiality of existence. The deficiency of origin also harbours a resource. It is certainly not a compensation, or a salvation, or a dialectical leap. But it is indeed the genius of the idiot.

Melancholy is perhaps not just black, or it may be that this blackness has a strange, elusive and yet unmistakable glow. In a sense, this does not surprise us: we are well aware that all of Stiegler's work is guided by the desire to act and by the will to believe or more precisely by this 'necessity of belief' that he takes from Kant as the most rational affirmation of reason itself. Reason therefore involves a default of knowing but it makes of this default its most powerful spring. '*Reason is a necessary default*', *un défaut qu'il faut*, writes Stiegler.[7]

The same logic governs the thought of the *who* (which is at bottom the general form of the idiot or the genius), since it presents itself without a past in its memory, not because it would have been forgotten but because it is yet to come. Here too there is a necessary default, an essential resource that comes from an in-principle lack (and the lack of a principle).

3

That melancholy contains an ambivalence, that the disastrous [*funeste*] rubs shoulders with the brilliant [*génial*], is confirmed in *The Age of Disruption*, where we find perhaps (to my knowledge at least) one of the last passages on this subject. It is, first of all, identified as the height of the 'loss of morale', which here refers to culture, where the latter must be cultivated and cared for, to avoid demoralization and the loss of meaning. It is tantamount to a 'loss of reason' that 'since Aristotle' has borne the name of melancholy.[8]

Just prior, the same phenomenon had been designated as that 'default of origin' that Heidegger named *Abgrund*, and that 'since ancient Greece has been called ὕβρις: excess, madness and crime'.[9] It is this that the 'the noetic soul *constitutes*' in its default of origin, and this is therefore also its 'fundamental risk'.[10] Let us note that if Heidegger does indeed refer to the *Abgrund* in *An Introduction to Metaphysics*, which is the example given here by Stiegler, it is however never in relation to hubris (generally speaking, this relationship does not seem to be significant for Heidegger). Certainly, Stiegler does not expressly connect them, but he does indeed posit an equivalence between the two terms. In other words, this minimal philological circumstance seems to betray a shift of value between *Abgrund*, which as Stiegler knows very well is the property of Heideggerian being, and 'excess, madness and crime', which is precisely what this book is asking how to escape.[11]

There would be some vacillation, here, between melancholy and madness – a fleeting vacillation, which might even seem to be due to a somewhat hasty dictation of this passage. But this is, of course, something of which we should take note, just as we have noted the somewhat odd use of the adjective '*funeste*'.

In any case, the possibility of hesitation and ambivalence at the edge of melancholy will be confirmed two pages later in the same book. Speaking of the 'neganthropological courage' that characterizes resistance to demoralization and its overturning, Bernard specifies that this courage that is 'neither denying nor repressing the disaster [...] does not sink into melancholy (even if it does not cease to challenge it and be tested by it, which it therefore does not deny)'.[12] The repetition of the verb 'deny' over the course of a few lines gives a double indication: on the one hand, the denial of the disaster that is happening to us would be analogous or parallel to the denial of melancholy and thus of the mortal default of origin; on the other hand, if we must deny denial on these two levels, it is because they cannot be far apart.

One might be tempted to deny melancholy, just as the disaster is denied by those who see only the continuous progress of transhumanism. But to deny melancholy one would have to deny death – this is also what transhumanism aims for, but as a possible future, whereas it is not possible to deny the default of origin to which the feeling of disaster [*sentiment funeste*] attests. One cannot escape the ordeal. One can, however, not sink into it but instead resist it, *even though* we are undergoing the ordeal of the irresistible.

What this involves stems from a parallelism between the coming disaster and the original loss of origin. This parallelism was introduced without being announced. It tends to turn into an identification, and this turning revolves around melancholy. The latter is not just the '*sentiment funeste*'. It is also that which is provoked by the disaster of the Anthropocene: the rupture of the 'feeling of existence' through which the 'moral being' is formed, as is said immediately after. What this now concerns, namely the courage of resistance, must therefore be just as possible – and demandable – in relation to the default of origin. This is indeed how we can understand how *it is a necessary default*. Or how the idiot can be a genius.

4

The situation thereby produced is as follows: at the point of melancholy there seems to intersect, without meeting, the irrepressible feeling of disaster and the feeling of existence. The second cannot deny the first. But what does it do with it? It is this question that seems to me to arise here. It imposes itself as the question of the operation or the operativity of negativity. If the default is necessary, what is the spring of this necessity?

It seems to me that Bernard is doing everything he can to set up this question, but that he does not answer it. I hasten to add that I am not pretending in this way to critique him. Firstly, this would require a much more meticulous journey through the texts. In this respect, however, it is notable that melancholy, despite its extreme importance, is not expressly problematized. It follows that it is quite possible that a flaw [*défaut*] in the argumentation or the analysis is necessary here for essential or transcendental reasons: such a fault is needed in order to pass from despondency to courage, from nihilism to confidence. In other words, a leap is needed, à la Kierkegaard, and not a continuous deduction. And if this is so, it is perhaps in this way, already, that the death of Bernard belongs to his work: as the moment of a passage to the act that all books, all conferences, all constructions of concepts could not accomplish. Moreover, this is what is indicated, towards the end of the book, by the need to approach the very impossibility of questioning.

What, then, is the spring of this necessity? I believe – and I say 'I believe' with the emphasis given to it by Bernard – that it can be found in negativity. The entire operation called negentropy, neganthropy, neganthropology is an operation of the negation of negation. No doubt it takes place in the recourse to the *pharmakon* and thus to the conjunction of contradictory aspects. And never would Stiegler allow it to be understood as a dialectic, since the possibility that it could lead to a final synthesis has been excluded. The double character of melancholy – the disastrous feeling experiencing also the feeling of existing – responds to the same exigency.

But make no mistake about Hegel. One should not believe that the negation of negation lies in an overhanging position. On the contrary, Hegel himself explains that the third moment is not a result but an uninterrupted movement. In what does this movement consist?

5

Paradoxically, it consists in a 'tarrying' [*séjour*]. The return in negation not only of what was denied, but also of its very negation engages the interminable movement. This is not said by Hegel only in these abstract terms. It is said also in that very famous passage of the Preface of *Phenomenology of Spirit* in which Hegel declares that the spirit does not retreat before death but maintains itself in it.[13] The Christological model is obvious: it is a matter of the two days spent by Christ in the tomb – a time about which nothing is said in the Scriptures (which nevertheless indicate its length) but which gives rise to the iconography of Christ dead in the tomb supported by angels. The Hegelian interpretation – which he develops in other texts – does not exactly fit the major theological interpretation (whatever the doctrine), although it corresponds to the hymn of Johannes Rist elsewhere cited by Hegel ('God Himself is dead')[14] and also, in a less expected way, to the commentary by Thomas Aquinas on Christ in the tomb: His death, he says, has not the semblance of night, but of day. This luminous death is no other death than death itself.

There is in Hegel a suspension of the dialectic on death and in death itself. The tarrying of the spirit is neither simply the negation of finitude and sin (and therefore of default) nor the negation of this negation as a restored or acquired infinitude. Or more precisely, if we bring together the texts of the *Logic* and the *Phenomenology*, the infinite to which the finite accedes is movement, infinite transformation, or else – the infinite present here, in place of death.

Yet this place is no less that of melancholy. Rist's hymn speaks of the sadness and sorrow of knowing the death of God. I want to suggest that Stiegler is not quite as far from Hegel as he thinks. Just as Hegel is the first thinker of an epoch that would be 'grey'[15] (let us translate: entropic, or 'without epoch'), so too it is he who introduces into philosophy a melancholy that we should not be too hasty to believe vanishes into the fulfilled Concept. On the contrary, the whole thought of which I have just sketched the outline is, we must say again, a thought of the sojourn, of tarrying: in some way, it is a matter of inhabiting melancholy.

Perhaps no thought since Hegel has been able to avoid this necessity. In this sense, I would like to say that Bernard comes as close as possible to this exigency. In this, too, he is doubtless a continuator of Derrida – but again, in other ways and according to another expectation.

6

His own way – it seems to me – is that of indecision about negativity. He is well aware of the negativity of the human event (with the melancholy that is its price), but when he sees it rushing headlong into self-destruction, he tries in his turn to deny it. Too astute to imagine overturning it into assured positivity, he affirms the impossibility (this is the whole end of *Disruption*) of its realization, claiming for it the status of dream or of a promise that must remain to come. Yet he also asserts that there must be a 'leap into a new era'.[16]

One can – or in any case I can – only follow him, and the same goes for when he speaks of 'sense arriving in non-sense'.[17] The fact remains, however, that the necessity of melancholy, as well as the necessity of not avoiding its ordeal, do not, strictly speaking, give rise in him to a reflection. Instead, what takes place is his brutal death, which in a single blow (it is indeed a matter of a blow) resounds for me like an echo of the Hegelian 'tarrying' of the spirit in death. Everything happens as if – regardless of the physical circumstances that overwhelmed him – Bernard had understood – or felt, experienced, here it is the same thing – that any attempt to make the leap into a new era *also* means leaping or welcoming a leap into the unthinkable, into the irremediable default, where no sense can arrive.

This is how his death belongs to his work, and this also means, immediately and urgently, that he entrusts us with a task. It is as if I heard him to be saying: no reversal of entropy (which, anyway, is that of the solar system: he talked about it in his first book) can avoid the melancholy sojourn. He said, 'melancholy is an experience of entropy by default'.[18] A very clear and decisive sentence, but at the same time enigmatic, since it talks about the experience by default of the default of origin. The experience of this default is impossible; yet, in the absence of making it possible, melancholy gives neither a representation nor an evocation but nevertheless an experience. There would thus be an experience of that of which there is no experience. In other words, an experience that I do not have [*je ne fais pas*] but that makes me [*me fait*] (as is the case with any true experience, which is not an act, but a suffering). Which thus makes me by undoing me (unravelling me, letting go of me and therefore of everything). In melancholy there would not be a presentiment or a fantasy of my death, but my death itself – something of the reality of my death.

What does this mean, beyond the observation, made by Bernard, of an experience or an ordeal [*épreuve*] that one cannot deny? It may mean nothing that unfolds into philosophical propositions. But it is an opening, and Bernard's death opened a meditation, that is, a contention (more than a retention and protention), one that is difficult, painful, risky, unmasterable but indispensable. A contention or a station, as the ascents of Sufi Masters are marked with stations. It is not enough to advance, to progress – even if it is a counter-progress: it is also necessary to stop, to experience the contention of a station, perhaps the time to listen to Paul Celan letting sense arrive and depart:

Down melancholy's rapids
past the blank
woundmirror:
there the forty
stripped lifetrees are rafted.

Single counter-
swimmer, you
count them, touch them
all.[19]

Translated by Daniel Ross

Notes

1 I will not give references because this is not an academic study. Moreover, the context always makes it possible to identify the source of the quotations. In addition, I will write 'Stiegler' or 'Bernard' according to whether I think or rather feel that it is a little more a matter of the author or of the man – and to mark that in him the two are not so clearly distinguished, as he himself notes in more than one text. *Translator's note*: Without at all wishing to affect the status of the text as something other than an academic study, the translator has added some references to assist readers who would like to follow up on the author's citations.

2 *Translator's note*: That is, page 141 of *La technique et le temps 1: La faute d'Épiméthée – 2: La désorientation – 3: Le temps du cinéma et la question du mal-être* suivis de

Le nouveau conflit des facultés et des fonctions dans l'Anthropocène; Bernard Stiegler, *Technics and Time, 1: The Fault of Epimetheus*, trans. Richard Beardsworth & George Collins, Stanford: Stanford University Press, 1998, 131.
3   *Translator's note*: Stiegler, *Technics and Time, 1*, 190, translation modified.
4   Stiegler, *La technique et le temps 1-2-3*, 8.
5   *Translator's note*: The usual spelling, and the spelling used in the title of Jackie Pigeaud's translation of Aristotle, is *mélancolie*. Cf. Jackie Pigeaud, *Aristote. L'homme de génie et la mélancolie: Problème XXX.1*, traduction présentation et notes de J. Pigeaud, Paris: Rivages, 1988.
6   Stiegler, *Technics and Time, 1*, 199.
7   Bernard Stiegler, *Technics and Time, 3: Cinematic Time and the Question of Malaise*, trans. Stephen Barker, Stanford: Stanford University Press, 2011, 182.
8   Bernard Stiegler, *The Age of Disruption: Technology and Madness in Computational Capitalism*, followed by *A conversation about Christianity with Alain Jugnon, Jean-Luc Nancy and Bernard Stiegler*, trans. Daniel Ross, Cambridge: Polity, 2019, 221.
9   Stiegler, *The Age of Disruption*, 221.
10 Stiegler, *The Age of Disruption*, 221.
11 *Translator's note*: The subtitle of *The Age of Disruption* is, in its original French edition, *How Not to Go Mad*.
12 Stiegler, *The Age of Disruption*, 224.
13 Georg W.F. Hegel, *Phenomenology of Spirit*, trans. Arnold V. Miller, Oxford: Oxford University Press, 1977, §32.
14 Georg W. F. Hegel, *Lectures on the Philosophy of Religion, Volume III: The Consummate Religion*, trans. Robert F. Brown et al., Oxford: Clarendon Press, 2007, 125.
15 Georg W. F. Hegel, *Elements of the Philosophy of Right*, trans. Hugh B. Nisbet, Cambridge: Cambridge University Press, 1991, 23.
16 Stiegler, *The Age of Disruption*, 308.
17 Stiegler, *The Age of Disruption*, 307.
18 Bernard Stiegler & Mehdi Belhaj Kacem, *Philosophies singulières: Conversation avec Michaël Crevoisier*, Paris-Zurich-Berlin: Diaphanes, 2021, 144.
19 Paul Celan, *Breathturn into Timestead: The Collected Later Poetry: A Bilingual Edition*, trans. & commentary Pierre Joris, New York: Farrar, Straus & Giroux, 2014, 88.

Chapter 4

JUST THIS, WRITTEN JUST HERE AND JUST NOW, BY JUST THIS INDIVIDUAL IN JUST THIS MOOD

Daniel Ross

> Every species of living being, and every specimen of each species, is affecting and modifying the biosphere by its efforts to keep itself alive during its brief lifetime. However, no pre-hominid species has ever had the power to dominate the biosphere or to wreck it. On the other hand, when a hominid chipped a stone with the intention of making it into a more serviceable tool, this historic act, performed perhaps two million years ago, made it certain that, one day, some species of some genus of the hominid family of primate mammals would not merely affect and modify the biosphere, but would hold the biosphere at its mercy.
>
> Arnold Toynbee, *Mankind and Mother Earth*

The turn towards curtailment

In the ecological sense (rather than the renewable energy sense), biomass is the sum total of all the living biological organisms occupying an ecosystem, while the ecosystem that encompasses all the smaller ecosystems has been known since the work of Vladimir Vernadsky as the biosphere. Vernadsky reflected on the way in which biochemistry interacted with and also transformed geochemical processes, hypothesizing that it was indeed possible, across geological timescales, for the combined biochemical processes of the biomass to reshape the whole 'terrestrial envelope'.[1] The book in which he made this case, *The Biosphere*, was published in 1926, the same year in which Martin Heidegger was still settling upon his conception of temporality while writing the final draft of *Being and Time*.[2] The latter was read by Bernard Stiegler early in his philosophical life, and his first book argues that, even though Heidegger does consider the place of artefacts in the constitution of Dasein's historical temporality, in the end he conceives all instruments (clocks, for example) as determining, and measuring, and making calculable, what exceeds determination, measurement and calculation, and so, according to Stiegler, Heidegger ultimately rejects the possibility that 'determining the undetermined' could grant access to the true character of time. For Stiegler himself, on the other hand, the possibility of such access can arise only from the world opened up by artefacts of all kinds, from the most basic tools to books to computer technology – the question of 'technics and time' lying in this possibility of a *world opened up* beyond the 'milieu' described by Jakob von Uexküll for the animal (or the sensible soul, in Aristotle's terms).[3]

The first of these books, *The Biosphere*, was not read by Stiegler until many years later, I believe, when he had begun to ask not just about technics but about what he called 'exosomatization', a notion derived from reading the work of Alfred Lotka, a mathematical biologist, on 'exosomatic evolution'.[4] Exosomatic evolution refers to the unfolding of a form of life that is no longer just the *endosomatic* evolution of the biological life of the biosphere, but rather technical life, that produces organs extending outside the body of the organism, without which it cannot survive. Exosomatic evolution thus names the process that in *Technics and Time, 1*[5] Stiegler mostly referred to as hominization (a process that *precedes* and constitutes the human).

What was crucial for Stiegler about the work of Lotka, however, was not just the *distinction* between exosomatic and endosomatic evolution, but the fact that these were considered to be distinct forms of the struggle against *entropy*: 'as was pointed out years ago by Boltzmann, the life struggle is primarily a competition for available energy'.[6] In other words, if, asking in Dublin in February 1943 about the fundamental character of life, Erwin Schrödinger referred to 'negative entropy', then Lotka shows in 1945 that we must ask this question in a specific way when life operates not just through natural selection, but artificial selection. Vernadsky himself would cite earlier work of Lotka, but whereas Vernadsky still believed that technology is a 'universal, peaceful and civilizing force',[7] Lotka, writing in the immediate aftermath of the Second World War, was far more conscious of the fact that, for exosomatic life, technical power (the 'receptors and effectors') requires processes of 'adjustment' through knowledge, wisdom and care, if the perpetual threat of disaster is to be avoided:

> It is precisely this that has gone awry in the schemes of men: the receptors and effectors have been perfected to a nicety; but the development of the adjustors has lagged so far behind, that the resultant of our efforts has actually been reversed. From the preservation of life, we have turned to the destruction of life; and from expansion of the human race we have, in some of the most advanced communities, turned to its curtailment.[8]

Stiegler, too, would characterize human existence in terms of this struggle between advance and delay, as being capable of leaps ahead but also of lagging behind, this corresponding as well to the 'intermittency' of the noetic soul according to Aristotle, that is, of the fact that this soul may be perpetually noetic in potential, but is noetic in actuality only at certain moments, and always at risk of falling back, both individually and collectively, a risk that is bound to occur, time and time again.

Arnold Toynbee and the enigmatic question

Another of the books that Stiegler chose to read seemingly towards the last few years of his life, since he referred to it only latterly in his work, is *Mankind and Mother Earth*, by the British historian Arnold Toynbee, written in 1973 but published posthumously in 1976. The life of Arnold Toynbee, from 14 April 1889 to 22 October 1975, occupies an almost identical stretch of time as that of Martin Heidegger, from 26 September 1889 to 26 May 1976. This span stretches from: (1) a youth that coincided with the industrialization of production and faith in 'progress' characteristic of the world prior to the unprecedented destruction of the First World War; to (2) a major period of work, a period that also saw the industrialization of consumption and the withering of faith in progress that accompanied the catastrophic convulsions of the twentieth century, and eventually to (3) late reflections, occurring at a moment when the 'glorious' post-war years were at an end, and the world was entering a new age of computational technology which by the 1970s had only begun to unfold into the vast process of transformation still underway in 2020.

In the third of these periods, Toynbee, like Lotka after the Second World War, had become highly conscious of the fact that humankind was now confronted with the reality of this curtailment, and the fact that the preservation of life had turned to destruction: 'Man is the first species of living being in our biosphere that has acquired the power to wreck the biosphere and, in wrecking it, to liquidate himself.'[9]

Toynbee does not discuss the second law of thermodynamics, and nor therefore does he discuss life in terms of the struggle against it, let alone exosomatic evolution. Nevertheless, the conception of the character of the 'mankind' of the title resonates with Stiegler's Promethean and *Epimethean* description of neotenic man in *Technics and Time, 1*. Toynbee writes about the process of hominization as follows:

> By the time that Man had become human, he had been stripped of all built-in physical weapons and armour, but he had acquired a conscious intellect which could think and plan, and two physical organs, his brain and his hands, which were the material instruments for his thinking, his planning, and his attempt to achieve his purposes by physical action.[10]

Furthermore, just as Friedrich Engels had said as early as 1883 that the hand 'implies the tool',[11] just as Stiegler will argue in 1994 that the evolution of this brain and this hand are co-constitutive with the evolution of the tool, so too Toynbee argues in 1973 that 'tools are coeval with human consciousness'.[12] Hence the situation with which we are confronted today, the power if not the likelihood that we will wreck the biosphere, is one whose roots lie so far back in time as to precede the appearance of *Homo sapiens* itself, in that coeval unfolding of hominization and technicization, of brain and hand, that gave rise to an exosomatic being who slowly but surely began to encroach further upon the biosphere, in the competition for available energy. Yet even if these roots lie far back, it is a transformation in the *conditions* of this competition that has produced a decisive turn from preservation to destruction:

> Since the beginning of the Upper Palaeolithic Age, perhaps 70,000/40,000 years ago, Man has been taking the offensive against the rest of the biosphere; but it is only since the beginning of the Industrial Revolution, no more than two hundred years ago, that Man has become decisively dominant. Within the last two centuries, Man has increased his material power to a degree at which he has become a menace to the biosphere's survival […].

Like Lotka, Toynbee sees this in terms of a lag in the adjustors, that is, a failure of knowledge, wisdom and care to keep up with this acceleration:

> […] he has become a menace to the biosphere's survival; but he has not increased his spiritual potentiality; the gap between this and his material power has consequently been widening; and this growing discrepancy is disconcerting; for an increase in Man's spiritual potentiality is now the only conceivable change in the constitution of the biosphere that can insure the biosphere – and, in the biosphere, Man himself – against being destroyed by a greed that is now armed with the ability to defeat its own intentions.[13]

Only an increase in spiritual potentiality can save us. If we do not wish to take this as a mystical invocation, then we must hear it as a diagnosis of current pathologies and a call to address them by changing the conditions of our psychic and collective formation. The exosomatic being may be in possession of a noetic soul, but there is no guarantee it will not be lost, in gaining the world (and it is for this reason that Stiegler will refer, in invoking and updating Weber's account, to the lost spirit of capitalism).

What then is the pathology? It is a recklessness, a carelessness that Toynbee identifies as a kind of civilizational suicidal tendency:

> Mankind's material power has now increased to a degree at which it could make the biosphere uninhabitable and will, in fact, produce this suicidal result within a foreseeable period of time if the human population of the globe does not now take prompt and vigorous concerted action.[14]

Forty-five years later, this prompt and vigorous concerted action has utterly failed to materialize, and we cannot discount that the moment may already have passed, that we have already slipped the hangman's noose we fashioned for ourselves around our neck, climbed atop the chair and are now teetering on the edge of the precipice, staring down into an abyss that can no longer be avoided. This is the point to which we have been brought by those two centuries that are now called the Anthropocene; it seems to be the point to which the Covid-19 pandemic is bringing us; it is the threshold that some noetic souls have already found themselves unable not to cross. As Toynbee says in the last lines of his book, in the case that we have just referred to as the Anthropocene, such a suicide cannot be divorced from a murder – on the scale of the biosphere:

> Will mankind murder Mother Earth or will he redeem her? He could murder her by misusing his increasing technological potency. Alternatively, he could redeem her by overcoming the suicidal, aggressive greed that, in all living creatures, including Man himself, has been the price of the Great Mother's gift of life. This is the enigmatic question which now confronts Man.[15]

Overshooting the mark, life itself

How should we characterize such a tendency towards civilizational suicide and biospheric murder? Is it not precisely the embodiment of the meaning of entropy, applied not just to thermodynamics or to biology (but incorporating both of them, and at the macrocosmic scale of the biosphere), but rather the entropy that is peculiarly characteristic of the kinds of beings that we ourselves are? In this tendency towards lagging behind, towards falling back from the noetic heights, towards destruction rather than preservation, towards disorganization rather than organization, towards the closing of what with so much difficulty had been opened, towards laziness and cowardice, hubris and denial, towards failing to do what one knows without a doubt must be done, towards suicide and towards murder, do we not see all of the elements that define the real meaning of this strange word designating *that turn within*, as it applies not just universally, but cosmically, and for us?

Such a question may sound like an affront to physics, as if philosophy has overshot the mark of its questions and landed in foreign territory over which it claims no rights. Nevertheless, all of these tendencies, taking the form of overreactions, underreactions, denials, suicides, murders and so on, can be considered as a kind of entropy, one peculiar to the exosomatic beings that we are and that Stiegler calls exorganisms. Such tendencies are irreducible and ineliminable, and the struggle against them is a temporary and local effort, even if the biological struggle of endosomatic evolution has lasted a few billion years within that largest locality that is the biosphere. This struggle to decrease entropy here and now can never occur without also producing an increase of entropy 'elsewhere', and, for the kinds of beings that we ourselves are, this struggle is not just against the second law of thermodynamics that is named by the concept of entropy, but against the peculiar regressive tendency it possesses that Stiegler calls anthropy. For Stiegler, the struggle for the preservation of exosomatic life is conducted under the sign of neganthropy, and the potential for unexpected turns and bifurcations this struggle contains, and must always contain as the *différance* we must *make*, relies on our anti-anthropic capacity.

At the same time, such tendencies, such reactions, such suicides and such murders – individual, collective, civilizational and biospheric – can be seen from a slightly different perspective: if the struggle against entropy is always a question of 'organization' (leading to Stiegler's distinction between the endosomatic organic and the exosomatic organological), and if this organization is always a question of strongly interrelated local and temporary dynamic systems (such as a cell, an organism, an ecosystem or an exorganism, a family, a tribe, a society, a civilization), then this organized locality always involves a relationship between interior and exterior that we can consider in terms of what in biology are known as immune systems. More than that, we can follow Georges Canguilhem in considering that there is no fundamental difference between a 'normal' immune response and a supersensitive immune response – such as the kind that produces a shock reaction that can even prove fatal to the organism itself, so that a protective mechanism in fact functions as a destructive agent. The difference is only in the *effects* of those tendencies in different contexts (different relationships between various scales of 'inside' and 'outside'), such that, for one and the same tendency, it may prove, for the locality in question, to be propulsive (leading it towards a future metastability) or repulsive (destabilizing the organism or exorganism).[16] Canguilhem himself reflects on the distinction that characterizes exosomatization when he refers to the 'technical form of life', but it is precisely *this* form of life that, he says, bears within it the '*temptation* to fall sick',[17] and hence we argue for the possibility of seeing the shocks and reactions of contemporary existence as tendencies towards a form of anaphylaxis that is not biological but *psychosocial*.

This affront thus risks being one not just against physics, but against biology and medicine too. Nevertheless, we have already seen Lotka citing Ludwig Boltzmann that the life struggle is a question of available energy, and we are inclined to extend the question of energy past the thermodynamic and to the libidinal, if not beyond. In terms of physics, what led to the discovery of the second law of thermodynamics was the inefficiency of the steam engine, and the effort to decrease that inefficiency through theoretical and practical means. From the consideration of the engine as a closed system, it was recognized that inefficiency itself is irreducible, and that *some* energy must always be lost by the system: it was the generalization of this realization to the universe considered as itself a closed system that led to the formulation of the concept of entropy.

At the time of Rudolf Clausius, however, atomic theory was not yet accepted, and so it was not possible to consider the entropic tendency of a gas, for example, in terms of the statistical consequences of vast numbers of atomic or molecular collisions. Only seven years later, however, in 1872, although the situation with regard to atomic theory had not changed, Ludwig Boltzmann would venture to formulate this law in statistical terms, which is to say, as the tendency for the probable to eliminate the improbable, which, over the vast numbers of such collisions involved in any macroscopic phenomenon, becomes an overwhelming and highly predictable progression that can also be understood as the elimination of the past. It is this characteristic of statistical dependability and this alone that makes this tendency susceptible to being described as a *law*.

How then did Boltzmann himself, who was a frequent lecturer willing to consider the relationship between scientific discovery and 'philosophical' questions, conceive biological life in relationship to this tendency? In an address on the second law given in 1886, a year before the birth of Schrödinger and three years before the births of Toynbee and Heidegger, Boltzmann recognized that all of terrestrial life depends on a constant source of energy from outside the biosphere, the Sun, allowing it to behave not as a closed system but as an open one. As a result, endosomatic life is, in sum, an effort to decrease and postpone the entropic tendency by taking advantage of this seemingly eternal source of 'free' energy:

> The general struggle for existence of animate beings is therefore not a struggle for raw materials – these, for organisms, are air, water and soil, all abundantly available – nor for energy which exists in plenty in any body in the form of heat [...], but a struggle for entropy, which becomes available through the transition of energy from the hot sun to the cold earth. In order to exploit this transition as much as possible, plants spread their immense surface of leaves and force the sun's energy, before it falls to the earth's temperature, to perform in ways as yet unexplored certain chemical syntheses of which no one in our laboratories has so far the least idea. The products of this chemical kitchen constitute the object of struggle of the animal world.[18]

The struggle for existence is a 'struggle for entropy', or, as Schrödinger will say, in a slightly less confusing way, more than half a century later: the organism 'feeds on negative entropy' to 'compensate the entropy it produces by living'.[19]

In 1905, almost twenty years after this statement about life and entropy, the chemistry of photosynthesis was still not understood, but Boltzmann would turn to the question of *value*, which is to say, the criteria of behavioural selection for exosomatic beings. While endosomatic beings are bound by instinct, or may partially modify it as a means of adapting to lessons learned in the life of an animal, and while the criteria for these bound selections are themselves the result of endosomatic evolution, this is no longer the case for exosomatic life. In the latter case, behavioural selection becomes a *problem* of existence, unbounded by the dictates of instinct even if it is still tied to the material reality of biology, and imposing the necessity of asking the question: what to do?

We are in the habit, Boltzmann says, of 'assessing everything as to its value', and of doing so 'according to whether it helps or hinders the conditions of life'. 'Life', whatever is named by this grand term, is the source of the judgement about values, and what has value is what promotes life. From here, however, Boltzmann takes a perhaps surprising turn, if not to say a twist: 'This becomes so habitual that we imagine we must ask ourselves whether life itself has a value.'[20]

What inner turn in the psychic individuation process of Ludwig Boltzmann produced this question?

Boltzmann intends to put before us evidence of a kind of perversion or distortion of thought. 'Metaphysics', having acquired the habit of pursuing questions and seeking out further territory to be conquered by those questions, applies that habit in ways that forget that the source of the possibility of such a habit is life itself qua archi-criterion of value. This problem does not arise for endosomatic life, which does not open up the gap into which the world enters and questions impose their necessity. Only for an exosomatic being, possessing a noetic soul that must *select* its behaviour, does it become possible to *ask* about 'value', to make life into a *question*, a question of value.

Yet only for such beings, too, does the possibility of an inner turn arise, an *entrope*, so to speak, allowing the question to rebound upon the source of its own unboundedness. It is the circuit through this outside that is the opening onto questions that introduces the possibility of this perversion or this distortion, which is inherent to questioning itself, and insofar as to question is *always* a question of responding to the problem of the exigencies of exosomatic life or the shocks it encounters or brings upon itself. It is a matter of the pharmacology of the question (in the sense of the Greek *pharmakon* that is a remedy that can always become a poison). And this is what Boltzmann himself understands, and why he raises this example:

> This becomes so habitual that we imagine we must ask ourselves whether life itself has a value. That is one of those questions utterly devoid of sense. Life itself we must accept as that which has value, and whether something else does can only be judged relatively to life, namely whether it is apt to promote life or not [and] this means whether life is apt to promote life, a question that has no sense.[21]

It is the characteristic of human beings to ask questions in general and questions of value in particular, to rise up above phenomena in order to see what chances life affords, and it is for this reason that Whitehead will say that the 'function of Reason is to promote the art of life'.[22] In asking the question of the value of life itself, however, in trying to measure that which provides the criteria for the measuring stick, it becomes a 'mental habit that overshoots the mark'.[23]

In fact, this trope of 'overshooting the mark' occurs repeatedly in Boltzmann's writing, like a habit of thought that occupies him constantly: the tendency for behaviours that begin as beneficial adaptations to be extended to an excessive point at which they become harmful. For example, in 1904:

> Many inappropriate features in the habits and behaviour of living beings are provoked by the fact that a mode of action that is appropriate in most cases becomes so habitual and second nature that it can no longer be relinquished if somewhere it ceases to be appropriate. I express this by saying that adaptation overshoots the mark. This happens especially often with mental habits and becomes a source of apparent contradictions between the laws of thought and the world, and between those laws themselves.[24]

What should regulate the relationship between thought and world is law, but the susceptibility of law to become a kind of automatism, repeated beyond all measure, leads to a kind of maladaptation, and this is precisely the case when, through a kind of repetition compulsion, we are drawn from asking about value within life to instead asking about the value *of* life, whether of the individual or of humankind:

> Similarly, something is called useful or valuable if it furthers the living conditions of the individual or of mankind, but we overshoot the mark if we ask for the value of life itself, when for example it seems to us pointless because it has no purpose outside itself. [...] I regard it as a central task of philosophy to give a clear account of the inappropriateness of this overshooting the mark on the part of our thinking habits.[25]

Yet it is 'metaphysics' itself that tends to fall into this trap, to overshoot the mark in a kind of excess that is also a kind of automatism – an excessive automatism. The task of philosophy must be precisely to aim at 'appropriateness' and eliminate 'inappropriateness' so as to approach a lawful expression without tangles and contradictions. Despite this seemingly straightforward, not to say naive commitment to philosophical perfection and the removal of all dogma, Boltzmann seems at the same time to recognize that a problem arises in this adoption of life as values of values, of life as the seemingly 'natural' archi-criterion of values, and that it is a problem, *precisely*, of the various scales of locality of this thing called life: 'In this we try of course to talk the individual into believing that what has value for him is not what promotes his own life but that of his family, tribe or even mankind as a whole.'[26]

What is the root of this need for persuasion? The implication, here, is that the values that promote the art of life and decrease the rate of entropy at least appear in different guises depending on the vantage point from which they are examined or sought, and more specifically depending on the scale of the locality with which one is preoccupied. Boltzmann seems to recognize that this unimpeachable value of values that is life 'itself', qua archi-criterion, involves in fact a composition of standpoints that is hard to reconcile with any kind of 'pure' reason – this merging of perspectives always requires a kind of stereoscopic trick or illusion. Is it not to effect this illusion, to pull off this trick, that the force of rhetoric is required?

Only such a trick or illusion of a depth that crosses scales can *guarantee* the unsocial sociability necessary to overcome an anti-social tendency towards preferring 'one's own' point of view, the wish to stay within the filter of one's own bubble. Yet without this guarantee, which always involves instruments to generate this depth and achieve this illusion, the social organism itself consumes itself entropically and anaphylactically. This is what Boltzmann seems to acknowledge here with his admission of the need for persuasion, but it is difficult to reconcile this apparent recognition of the need to conduct such a composition, to generate an always partly illusory or fictitious depth of field capable of crossing the scales of locality, with Boltzmann's immediate conclusions concerning the task of philosophy in the aftermath of the second law of thermodynamics, which seem premised, instead, on the possibility of eliminating this aspectual or perspectival implication:

> The task of philosophy for the future is, in my view, to formulate the fundamental concepts in such a way that in all cases we obtain as precise instructions as possible for appropriate interventions in the world of phenomena. This requires first that if we follow different paths, we never reach different rules for further thought and action, that is we never meet internal inconsistencies [...]. That sort of event is always a sign that the laws of thought still lack the last finish.[27]

Surely what we know from the second law of thermodynamics is that no system ever receives that last finish, that there is only entropic becoming or unfinished individuation.

Boltzmann gave many such lectures and lecture series in the latter part of his life. Himself a 'democratic radical and a resigned republican',[28] Boltzmann lived within the contradictions of Robert Musil's Kakanic Vienna,[29] and, faced with hostility towards atomic theory, especially from those around Ernst Mach, Boltzmann's initial taste of success turned to pain, anxiety and restlessness, and a powerful urge to convey, to anyone who would listen, the importance of his theories, ideas and discoveries – which were, after all, destined to be proven correct. Cuts in finances, health problems, overwork, swings between mania and depression, thirty lectures in English at Berkeley in 1905 that his listeners struggled to understand – all this led Boltzmann in the latter part of 1905 to visit a mental hospital, but he decided against staying. Mach (who was the subject of Musil's PhD) informs us that there had been earlier attempts at suicide, and it was understood that he required constant supervision. Boltzmann was forced to postpone his lectures of summer 1906 due to 'his nervous condition', but then travelled to Duino with his wife and daughter: apparently it was his wife's idea, but whatever the case, he seemed to improve.

As I am writing these lines, I realize with a sudden sense of shock that today is 5 September 2020, precisely one month after the death of Bernard Stiegler on 5 August 2020 at sixty-eight years old, and precisely 114 years to the day, after 5 September 1906, when Ludwig Boltzmann, aged sixty-two, due to return to Vienna the next day to commence his lectures, and apparently having 'showed himself particularly excited' earlier on that day, made and carried out the decision to hang himself.

A final birth? The last escape?

Among those who would be affected by this event was Erwin Schrödinger, then nineteen years old:

> The old Vienna Institute, from which shortly before Ludwig Boltzmann had been torn away in a tragic fashion [...], engendered in me a direct empathy for the ideas of that powerful spirit. For me his range of ideas played the role of a scientific young love, and no other has ever again held me so spellbound.[30]

The budding physicist, registering the shock of the tragic, not to say violent loss of the voice that had spoken to him like no other, the one who, *binding* him with the spells that were his ideas, incantations that had opened the enchanted realm that would become Schrödinger's discipline, would, by adopting this accident and striving to make it his necessity, participate in the transindividuation of the significance of the life and work of one who, for him, amounted, we could say, to a kind of saint – however improbably. Almost forty years later, in lectures given in Dublin, Schrödinger would, as we know, postulate the thesis that life must be considered a form of 'negative entropy'.

At some point in the last few years of his life, Boltzmann had begun a poem with the following lines:

> With torment that I'd rather not recall
> My soul at last escaped my mortal body.
> Ascent through space! What happy floating
> For one who suffered such distress and pain.

It is difficult not to hear in these lines the sense of mortality as a release from suffering yearned for by the distressed and unstable thinker of entropy as the statistically guaranteed triumph of the probable. Yet later in the poem, we read the following:

> The saint who suffers pain and grief
> Redemption's rays illuminate his way.
> No man achieves a hero's worldly fame
> Who has not forced himself with all his power;
> And as it caused his aching heart to tremble
> His valiant deed will live in song immortal.[31]

Only a few years later, in his *Duino Elegies*, Rainer Maria Rilke would refer in similar terms to the *kléos* of the hero:

> Begin again. Try out your impotent praise again;
> think about the hero who lives on: even his fall
> was only an excuse for another life, a final birth.[32]

A 'final birth', *seine letzte Geburt*, that lives on 'in song immortal'. Gilbert Simondon, in his own reflection upon the fate of the noetic soul of the mortal, will say a 'second birth', *une seconde naissance*.[33] What does it mean to say that there is a second birth that survives the first death? In fact, this involves the same question that animated Schrödinger's consideration of the animate as negentropic, the same question that Lotka asks concerning the doubling of the detour within entropy that occurs with exosomatic evolution, and which is to say, as well, the question just barely but already opened up by Boltzmann concerning the shifts of scale involved in the consideration of 'life' as the source of neganthropic value.

For Simondon, the second birth becomes possible when 'life' is no longer just a matter of vital individuation but of psychic and collective individuation. The psychic individual belongs not just to a vital milieu, an ecosystem with which it negotiates its metastability until the moment of its eventual demise, but rather exists in relation to other psychic individuals, and, more than that, exists *only* in relation to other individuals. Already while still living, after the first birth, the individual is both drawing from the collective and *secreting* himself or herself *into* the collective, through which alone the individual has the chance of becoming the one who they are or will be. The individual remains unfinished in life, and *continues* to be unfinished even after death, so long as there are psychic and collective individuals that remain affected by the life of the lost individual, by the work of the individual, if not as well by the shock of losing them, and who, so affected, carry something on – a survival. As the individual is withdrawn from life, that absence becomes the gap that opens the call to the work of 're-actualizing this active absence' as a 'seed of consciousness and of action'.[34]

Yet for all that to be possible, the remains are definitive, less the corpse than the corpus, the traces, through which what has been secreted may become the transindividuation of meaning, of significance – significations that become, precisely, transindividual. As a result, what has closed remains open, 'not contained, locked, in an individual enclosure that will degrade'.[35] Nevertheless, this also means: 'converted into signification, perpetuated in information', or in other words, exposed to the risk that, entered into a song immortal (and here we must not forget that a song, whether mortal or immortal, is always something technical, as is the act of singing), the significance of that song can fade away, not disappearing but awaiting reawakening and reactivation by a new singer, who inevitably confronts the same risk. Only in this way can the instability of the fragile individuation of a noetic soul be overcome and join the 'definitive and only metastability', that of the collective, 'perpetuated without aging throughout successive individuations'.[36]

Nevertheless, we would like to introduce two wrinkles into the tapestry we are here weaving between Toynbee, Boltzmann, Rilke, Schrödinger and Simondon (not to mention Lotka and Stiegler), two discordant notes into this cosmic melody concerning the posterity of the mortal individual sung by the chorus of the collective, amounting in the final assessment to an unending 'ode to man', and to his *deinon*.

The first is to suggest that, sometimes, the existence of a psychic individual is not only cut short, but cut short having left traces and memories of unfulfilled promise, of work yet to be carried out, of leaps yet to be taken, but indicated, prepared-for, initiated, promised. Secretions remain, unextruded and no longer extrudable. Sometimes those promises, unfulfilled, are of a singular kind utterly dependent on the unique characteristics of the particular soul who conceived possibilities and bifurcations that were not yet brought to fruition – in the case of genius, for example. No doubt continuation, a second birth, is still possible, and even more so, yet in another direction, a turn off and probably a turn in, and where the collective, insofar as it is a genuine collective, cannot but be aware of how the *direction* in which that living on continues involves a deviation from the initial path, and, perhaps, in a weaker direction, as a shadow.

The second concerns Simondon's thought that it is the collective alone, constituted through the immortal song of transindividual meaning, perpetually becoming signification perpetually in need of reactivation, that ever and always bears any lasting metastability. Maybe so, undeniably even, but if this song is technical, and this technics is a necessary medium and support through which alone the transindividual is possible, then we are confronted with what Stiegler calls a *pharmakon*, and with the fact that the collective individuation process opened up through this song need not just progress, but is always and irreducibly exposed to the risk of regress, just like the psychic individuation processes of which it is necessarily composed. The collective soul, too, is only ever intermittently capable of its noetic potential, and its greater stability, or the advent of a new reactivation, is by no means divinely guaranteed.

In short, while Simondon speaks as though, by being translated into transindividual signification and meaning, the psychic individual is being absorbed or reabsorbed into something larger, something of a broader amplitude, not a universe but a cosmos, nevertheless it may be that scales become skewed and reversed, that what is larger suddenly becomes smaller, and that the smaller scale of the psychic soul suddenly proves to be too vast for any such reabsorption. The whole world may shrink, and shrink before the task given to it by a loss that, precisely, exposes its shrunken character. In such a case, when the collective has suffered a proletarianization and denoetization (to use Stiegler's terms) that leaves it bereft and adrift, then the adoption of the shock of the event may not just be the improbable necessity of quasi-causally turning the accidental into the necessary: it may have become a strict and irreversible impossibility. Whether overshooting the mark in overreactions, or underreactively failing to fire a shot for want of ammunition or will, it may turn out that there is no longer anyone left to measure up to the task of thinking and caring in the Anthropocene. In the struggle between the upward trend and the downward, as Whitehead calls it, and so long as entropy remains an irreducible tendency of the universe, inevitably it eventuates that we will, one day, reach the cosmic inflection point.

What is the Anthropocene? What is it, futurally, but the arrival of a moment when nihilism is fulfilled, when there is no longer any way to 'talk the individual into believing that what has value for him is not what promotes his own life but that of his family, tribe or even mankind as a whole' – or the biosphere as a whole? And when this is so because there is no longer any voice, speaking or singing, capable of resonating with a sense of the scale of the problem that exosomatic life in the biosphere has become, no signification immortal or mortal capable of weighing the gravity of the task.

It is possible, after all, for the last such voice to be silenced, in this Anthropocene whose probabilistic and entropic character Boltzmann was the first to open up mathematically, and who, perhaps, was also the first to perceive this epoch's permanent and impending closing, foreshadowed by his own dramatic spiralling in upon himself. Other voices, too, have opened up our awareness of the general character of entropy, for instance philosophically, or organologically, calling us to the task of healing this surreal cosmos, yet they too can close, perhaps are closed, perhaps once and for all. Can we still dream that they might live on, a second birth? What else than a dream could they ever have been? It is this conjunction of closures, psychic, collective and biospheric, that, more than anything, confronts us with the appalling and dreadful likelihood of the Anthropocenic and anaphylactic fate that beckons us – the last escape.

Notes

1. Vladimir I. Vernadsky, *The Biosphere*, trans. David B. Langmuir, New York: Copernicus, 1998, 118.
2. Martin Heidegger, *Being and Time*, trans. John Macquarrie & Edward Robinson, New York: Harper and Row Publishers, 2008.
3. Jakob von Uexküll, *A Foray into the World of Animals and Humans, with, A Theory of Meaning*, trans. Joseph D. O'Neil, Minneapolis and London: University of Minnesota Press, 2010.
4. Alfred J. Lotka, 'The Law of Evolution as a Maximal Principle', *Human Biology* 17/3 (1945): 167–94.
5. Bernard Stiegler, *Technics and Time, 1: The Fault of Epimetheus*, trans. Richard Beardsworth & George Collins, Stanford: Stanford University Press, 1998.
6. Lotka, 'The Law of Evolution as a Maximal Principle', 179.
7. Jacques Grinevald, 'Introduction: The Invisibility of the Vernadskian Revolution', in Vernadsky, *The Biosphere*, 20–32, 25.
8. Lotka, 'The Law of Evolution as a Maximal Principle', 192–3.
9. Arnold Toynbee, *Mankind and Mother Earth*, Oxford: Oxford University Press, 1976, 17.
10. Toynbee, *Mankind and Mother Earth*, 575.
11. Friedrich Engels, *Dialectics of Nature*, trans. Clemens Dutt, London: Lawrence and Wishart, 1940, 17.
12. Toynbee, *Mankind and Mother Earth*, 17.
13. Toynbee, *Mankind and Mother Earth*, 17.
14. Toynbee, *Mankind and Mother Earth*, 17.
15. Toynbee, *Mankind and Mother Earth*, 596.
16. Georges Canguilhem, *The Normal and the Pathological*, trans. Carolyn R. Fawcett, with Robert S. Cohen, New York: Zone Books, 1991, 205.
17. Canguilhem, *The Normal and the Pathological*, 200, my italics.
18. Ludwig Boltzmann, 'The Second Law of Thermodynamics (1886)', in Ludwig Boltzmann, *Theoretical Physics and Philosophical Problems*, ed. Brian McGuinness, trans. Paul Foulkes, Dordrecht and Boston: Reidel, 1974, 13–32, 24.
19. Erwin Schrödinger, 'What Is Life?', in Erwin Schrödinger, *What Is Life?, with Mind and Matter and Autobiographical Sketches*, Cambridge: Cambridge University Press, 1992, 1–90, 73.

20 Boltzmann, 'On a Thesis of Schopenhauer's (1905)', in Boltzmann, *Theoretical Physics and Philosophical Problems*, 185–98, 196.
21 Boltzmann, 'On a Thesis of Schopenhauer's (1905)', 196.
22 Alfred North Whitehead, *The Function of Reason*, Princeton: Princeton University Press, 1929, 2.
23 Boltzmann, 'On a Thesis of Schopenhauer's (1905)', 197.
24 Boltzmann, 'On Statistical Mechanics (1904)', in Boltzmann, *Theoretical Physics and Philosophical Problems*, 159–72, 166.
25 Boltzmann, 'On Statistical Mechanics (1904)', 166–7.
26 Boltzmann, 'On a Thesis of Schopenhauer's (1905)', 196.
27 Boltzmann, 'On a Thesis of Schopenhauer's (1905)', 197.
28 Carlo Cercignani, *Ludwig Boltzmann: The Man Who Trusted Atoms*, Oxford: Oxford University Press, 1998, 43.
29 And Musil did not hesitate to interpret this Kakania in terms of the question of entropic becoming, and in dialogue with Nietzsche and his concept of nihilism, as shown in Bernard Stiegler, *Qu'appelle-t-on panser? 1: L'immense régression*, Paris: Les Liens qui Libèrent, 2018.
30 Erwin Schrödinger, quoted in Cercignani, *Ludwig Boltzmann*, 36.
31 Boltzmann, quoted in Cercignani, *Ludwig Boltzmann*, 46–8.
32 Rainer Maria Rilke, *Duino Elegies and The Sonnets to Orpheus*, trans. Alfred Poulin, Jr., Boston: Houghton Mifflin, 1977, 7.
33 Georges Simondon, *Individuation in Light of Notions of Form and Information*, trans. Taylor Adkins, Minneapolis and London: University of Minnesota Press, 2020, 240. For the following discussion of Simondon, the author would like to sincerely thank Anne Alombert, for drawing attention to these parts of Simondon's work, for her ongoing discussions concerning these questions, and for her thesis on Derrida and Simondon that greatly adds to the sum of our knowledge of these two thinkers by, for the first time, bringing them together in a mutually illuminating manner, thereby increasing the life of their second birth.
34 Simondon, *Individuation in Light of Notions of Form and Information*, 276.
35 Simondon, *Individuation in Light of Notions of Form and Information*, 240, translation modified.
36 Simondon, *Individuation in Light of Notions of Form and Information*, 240–1.

Part II

INTERMITTENCE: CARING TO BELIEVE

Chapter 5

CARE AS INVENTION: A TRIBUTE TO BERNARD STIEGLER

Anaïs Nony

To Stiegler's notion of *pansable* (curable), a word that recalls the *différance* so dear to his mentor Jacques Derrida, one might also need to add that *penser* (to think) relates to the Latin *penso*, the frequentative of *pendo*, to hang, suspend. The *pansable* (that which can be healed) is as much the *pensable* (that which can be thought) and the *suspensible* (that which can be hung). Stiegler's final act revealed that which was always already there: an unhealed pharmacological shadow that preceded him.[1] While he entered philosophy with the argument of technics as the *impensé* (unthought) of continental philosophy,[2] he concluded in a final acting out, an *impansable* (uncurable) that ended his life. He, who believed that life is about cultivating *rêveries* and protentions capable of promise.[3] Protention means both a capacity to invent and an ability to project oneself into the future through the practice of imagination and desire. He, who pondered about the retentions we have and the various forms of memory they take, and how to make them become the true modes of being-in-the-world.[4] I know now that his fatal transgression is as much an accident as a departure, an emotional ceasefire and a bifurcative ending. Somehow, he found a way to remain faithful to the originary beginning of his thinking *in act*.

His last act interrupted a long and painful series of repetitions. Stiegler referred to his first accident, the one that forced him to spend five years in jail, as a social suicide. He saw in his armed robbery a form of suicide that not only caused him to be incarcerated but forced him to investigate such a separation from the social world as the necessary default that pushed him to invent a new technique of self-care. Such a social suicide gave the impetus for thinking about the disaffection of individuals in societies where capitalized power has become the rule of law. During his incarcerated time, he was in his cell like a fish out of the water. It created an *epochē*, a suspension, and invented a new relation to this locality through discipline.[5] The cell required him to produce new realms of signification in order not to go crazy.[6] A few decades later, the *epochal* dimension of thinking one's condition together with the local dimension of relating to one's immediate environment found its uncanny symbol in the flying fish, the symbol of Stiegler's school.

In the school *Pharmakon.fr* launched in 2010, the flying fish is the allegory of the noetic soul, a spirit intermittently suspended above water, enacting forms of distant and associative gathering to transform the milieu in which it evolves. Through the intermittence embodied in the flying fish, Stiegler highlighted both the condition of *noēsis* and its function in life.[7] As a condition, intermittence allows the individual to cultivate a phasic relationship to oneself, what Simondon calls a *déphasage* and what Stiegler calls a bifurcation. The function of *déphasage* consists in the opening of a mode of becoming that is both processual (at the level of the structure) and phasal (at the level of its operation). *Déphasage* and *bifurcation* imply the restructuring of operation in the process of transindividuation. During his seminars and conference presentations, one could see a specific shape coming handy to explain such a transindividuating process. Stiegler used the shape of the spiral to show how an individual reactivates their becoming by taking a leap, by bifurcating, and thus reinventing forms of both psychic and collective belonging. The process of transindividuation operates as a structural suspension that activates new phases of being, new modes of existence. The spiral helped visualize alternative forms of becoming: forms that escape today's increased codifications, linearities and abstractions central to computational capitalism. To him, transindividuation was about the improbable bifurcation that an individual cultivates to transform her/himself. This bifurcation stood against the pervasive probabilities of algorithmically run capitalist automation.[8] In this context, the discretization of space and time by the digital challenges the process of individuation.

To Bernard the tension between the different temporalities of the discrete and the processual create the metastable milieu suitable for a new critique of our cultural and political condition. Today capitalism is based on the mathematical and industrial exploitation of the drives and the mimetic atavisms that underlie them.[9] To overcome the flattening and synchronizing tendencies of what he later calls *soft* or *fluid* capitalism, Stiegler's philosophy embraced a 'rhythmic attitude'[10] that is an intermittent force, one that cultivates rhythms of localities.[11] This force admits that an individual is a being in tension: their capacity to act, care, think, heal, share is always fluctuating according to ever-changing sets of internal and external limitations. I see the rhythmic intermittence embodied in the symbol of Stiegler's school as the condition for the development of anything that gives life a flavour, that which makes life worth living.[12] The intermittence of the noetic living being is a complex system in tension between process and phase. Bifurcation intervenes in this system by opening signification the possibility to emerge from the transductive and complementary mediation between operation and structure.

In a sense, and because the human was never a granted category, Stiegler's philosophy implied a humility regarding the limit of mankind. To him, the human (he would say the non-inhuman) is a noetic living subject defined by the invention of technique to supplement its originary *défaut*.[13] Embracing the *défaut* as much as the promise of our human condition is the force of Stiegler's philosophical project: one that requires not to neglect our lacuna and not to be lured by the ideological promise of technological determination. In other words, it is a project dictated by and dedicated to attentional forms of both psychic and collective care, a project that understood desire as the intensification of individuation.[14] I believe that, in Stiegler's philosophy, care is both a healing and inventing category. It operates from the retentional trace of the past (the *pansable*) and the protentional field of the future (the *suspensible*). Care as invention transforms the *impansé* (uncured) into the *pensable* (that which can be thought). For Stiegler, our intermittent capabilities are supplemented by our desire to act and take care. It is in the organology of desire for a collective future that the late philosopher attempted to take care, by paying attention to the disruptive forces that finally pushed him to take his own life.

Suicide is an act of transvaluation, inasmuch as it implies acting out according to a new belief that is both a secret wish and an act of care. As such, it is important not to neglect the suicidal dimension of his departure because of the transvaluation that such an act represents. For someone who spent most of his life questioning the pharmacological dimension of technique it would be a mistake not to include his departure as the symptom of a malaise. Such a malaise is a *mal-être*, a profound pain that functioned as a shadow in his work, intermittently feeding the strength and the fear of not doing enough to cure the societies we live in. I would then like to write about the memories I have and share the improbable protentions (as dreams and desires) we might have the audacity to invent in a collective gesture of thinking (*penser*), healing (*panser*) and suspending (*pendre*) the modes of existence of the spirit or *nous*, that inhabits us all. Such spirit requires thinking, healing and suspending outdated forms of belief, structure of discourse and collective symbols of hegemonic power in order to invent new modes of belonging.

The *nous* is the *spirit understood as the care taken of the objects and subjects of individual and collective desire*.[15] Care, spirit, desire and *nous* are the operative pillars of noeticity, that is, the spirituality which is life in potential, and which requires relational localities of care to be actualized. As such, *nous* is that which gives direction to our being, that which gives meaning and sense of purpose to our life. Purpose, sociality and orientation are increasingly lost in late capitalist societies. *Nous* refers to the complexity of the relationship of the spirit with itself; it is both a pre-individual and a trans-individual force and as such a mode of sensing the world and of belonging to it intuitively.[16]

Contrary to the increasingly cybernetic governance of our living practices, *nous* is a pilot that is not ruled by computational machines and technologies of control, but by techniques of care. Only care, as an intersectional practice of overcoming hegemonic power, can guide the making of new significations and new singularities. Caring is inventing what matters in the present to significantly shape the future. Caring means inventing a relational modality of belonging otherwise in the world: it is giving significations to things so they can matter, materialize, be actual and acknowledged as such. Caretaking is nootechnic[17] as it is psychic and collective: it finds agency in the technics we forge in relation to localities and remains a guiding force to navigate amongst increasingly disruptive technologies of power. In other words, care is an inventive category of knowledge (*savoir, savoir-faire, savoir-être*) that takes the spirit as its main axiom of becoming.

For Stiegler, a society is a system of care that relies on the production of attention.[18] *Attendere*, to shift one's attention, is to take care. To be careful, to be attentive and to care for something or someone is to pay attention. Yet, since the second tome of Bernard Stiegler's *La technique et le temps*[19] the late philosopher warned against the disorienting and disrupting effects of technological advancement, underlining the procedures and conditions of programmatology (as seen in calendars, cardinality and synchronized structures of tertiary memory). This technological advancement has created systemic delays mainly provoked by newly engendered technogenetic structures onto traditionally anchored sociogenetic operations. In other words, technogenesis structurally pre-empts forms of sociogenetic relations.

Let me recall a memory I have of the summer schools of the *École de Philosophie d'Épineuil-le-Fleuriel*. Since 2010, our conversations were based on the shared understanding that the place where dreams are shaped is an organological scene that calls for a taking care of our capabilities to project ourselves beyond our current environment. Hence the symbol of the flying fish, always enacting various bifurcative forms of belonging. During these summer schools, knowledge was the transitional object of our infinite discussions. Once a year, a very small village in the middle of France's countryside was populated by activists, artists and intellectuals who could dream to change the world, even if it meant embracing the possible limits of such a wish. We gathered to jointly create protentions, understood as forms of collective desire for the future.[20] These moments offered a relational mode of building common *savoirs/saveurs* based on desire as a product of social bonds. These bonds inhabited our souls long after the end of this utopian school.

In 2014, our meeting had a welcoming message: '*sogno dunque sono*', I dream therefore I am, and aimed to rethink the oneiric condition necessary for the production of knowledge. Knowledge was understood in the sense of *sapere* as that which accounts for the flavour of life. With '*sogno dunque sono*' Stiegler gave Descartes's *cogito ergo sum* a twist and invited his guests to collectively invent the stage from which one creates protentions. The task was to interrogate how computational capitalism digitally implements disruptive tendencies (as seen in an individual's lack of trust in themselves and others, the spread of misinformation, uniformity of behaviour, increased fascism, impunity, etc.). The goal was to address the way in which algorithmic governmentality was not only governing conducts but fundamentally shaping how law, rules and norms breed obedience.[21] Our concern was that protentions, which are produced by our capacity to suspend outdated beliefs and stereotypes, were being hijacked by the operative realm of everyday computational machines. The disruptive dimension of technologically driven societies was central to our interrogations and to our will to somehow, at our own scale, change the world. As such, the challenge was to address the tendential fall of our affective capacity for taking care of ourselves and others in a milieu driven by disruptive technologies.

That year, Kant's critique of the faculty of judgement was re-evaluated in light of Plato's understanding of knowing (*connaître*) as always being an acknowledging (*reconnaître*). Our reading of the Socratic dialogue *Meno* helped highlight that to develop knowledge (*savoir*) a certain disposition is required to cultivate, with care, that which gives flavour to our understanding. Such a cultivation of the place of knowing is the activation of what Simondon calls the transindividual, namely the relation between both a psychic and a collective individual as well as the relation between non-individuated realities, or pre-individual realities, within the individuated subject.[22]

What inspired me the most in 2014 was the understanding that the transindividual is made of spiritual realms of affective exchange. This spiritual realm, as I understand it, is both a process of transindividuation that highlights the operative circuits between the I and the we, between what Stiegler calls traumatypes and stereotypes,[23] and a mediation between pre-individuated and individuated beings. Understanding the principle of individuation as mediation and the circuits or relations that operate in the transindividual, allowed to discover *anamnēsis*, that which is produced in a dialogue with oneself or with the other, leading to a reminiscence of knowledge. In a world where dominant technologies implement necrotic programmes, the capacity to remember and to recall seemed as urgent as the capacity to suspend disruptive operations. Somehow, by focusing on the transindividual dimension of knowledge, I found a way to remember and recall, much like I am doing today as I type these words.

In 2014, dreaming replaced thinking as the condition and mode of *existing individually* but also *consisting collectively*. Dreaming became both a practice in collective envisioning and a tool in anti-entropic knowledge making. Much like Jonathan Beller, Jonathan Crary and Don Ihde, Stiegler warned us against the pervasive tendency of imaging technologies[24] which infiltrate the most intimate space of thought, hence shaping the organicity of machines, as well as the machinic organization of human beings.[25] For Stiegler, dreams are not just sleep and what is at stake is not simply the end of it.[26] Dreams are a condition of possibility for noetic life: the dignity of living according to non-inhuman conditions.[27]

In the system of Stiegler's school of philosophy, the spiritual economy of the people gathered in time and space facilitated the production of knowledge, with respect and dignity. The act of dreaming was psychic, collective and nootechnical: it aimed to be realizable according to the conditions of sufficient reason but also according to relations of force that are political, economic and ecological.[28] Imagination and interpretation became central categories of knowing, as we adopted Kant's transcendental imagination in Stiegler's organological interpretation of it, taking into account the technical exteriorization, while we updated it with the visionary invention of Simondon's cycle of image formation.[29] Together, imagination and interpretation became categories of the individual and collective desire to know and thus to take care.

The anthropotechnic capacity of knowing, central to Stiegler's philosophical project, is anchored in an *oneirology*.[30] Dreaming is a radical modality of engaging the increasingly entropic dimension of the world, it is an act of negentropic valuation. Because a technical life is a noetic life that realizes its dreams,[31] the intermittence central to the oneiric condition is the driving force of an epochal mediation which refuses the short-circuiting of societies imposed by disruptive technologies of machinic governance. Only a new noetic dream can become the epochal mediation necessary to interrupt and transvaluate the economic and ecological order imposed by machinic computation. Here, I cannot but think of Bernard's suicide and the last noetic dream he might have had the audacity to envision as he interrupted his own life.

The memories I have of the summer schools, the seminars and other adventures we created with Bernard since 2010 are intermittently inhabiting the knowledge I continue to forge and cultivate. This knowledge as memory is the yet-to-be-remembered as the always-already-known. In light of Stiegler's departure, the *pharmakogenesis* of care is a practice in anticipating what needs attention and requires healing. Much like individuation, healing occurs in a spiral where categories of time, space and their correlative speeds function via a different regime of expansion, extension, tension and relief. What needs healing imposes a temporality that cannot be subjugated to other priorities. While care is invention, that is, an anticipation of what needs attention – we could say a form of protention as much as a protection – healing is an intervention that is an act of repair and repatriation. Care is to healing what the transindividual operation is to the structure of individuation: an inventing force that adopts new bifurcative modes of existence.

If nothing else, the Anthropocene is a time when earth claims that it can no longer thrive in its *puissance* without collective effort for healing. If caring comes short, healing is required. The taking care (*être attentif*) of our healing practices (*pratiques de soin*) resides in a modality of being that is fundamentally transgenerational and ancestorial. The layers of self we peel off to further engage in a history that matters, reveal the intricacy and connectedness of both living entities and disruptive systems of power. This relation appears as both a correlative dimension and structure of being-in-tension and being intentional about the caring and healing values one cultivates. The affective process of healing takes for its anchor the formative dimension of taking care, one that shapes the phenomenon that needs attention. The ontogenesis of care is the genesis of the active investment in caring about the non-inhuman and inhuman dimension of life. In other words, I believe that care is invention: a relational mediation that cultivates resonances endlessly shaping the system where the operation and structure of individuation take place. As such, care is as much an economic and political issue as it is a psychic and collective challenge.

To think about the ontogenesis of care is to ask about the genesis of thought, projection, wishes, desires ... for care has been the forgotten notion of a continental philosophy rather busy with questions of power over dichotomies (body/soul, form/matter, inside/outside). The ontogenesis of care implies an organology of dreams, wishes and desires and as such it stands as a spiritual economy.[32] Such an economy requires a critique of both psychic and collective forms of attention and needs the development of nootechnics. Nootechnics takes the digital as an amplifying structure of reticulation, meaning that digital technologies must develop processes in which new relational modalities toward technics are deployed.[33] This reticulation defines the temporality of individuation and is based on the expansion, sharing and care of transductive unities. As a vital operation of reticulation, transduction needs to be addressed from different points of view: institutions, governments and power dynamics of modulation as seen in racial and gender-based discrimination.[34] In this context, the governance of memory, behaviour and invention is key to understanding our political and cultural condition in an era of anthropogenic disruption, an era that annihilates capacities and the necessary intermittent fluctuations of operations of thoughts and long-term circuits of investment.

Finally, if 'the reality of disruption is the loss of reason',[35] the reality of the Anthropocene is the loss of a common noetic dream capable of transvaluating our current condition. This noetic dream is not one that resists, but a dream which can overcome disruptive energies and their system of oppression. That is certainly one of the most difficult lessons of Bernard's departure. He might have known it from the beginning, but he dreamed of a *différance* and tried most of his life to enact it, to actualize it in various modes of existence. What we know now, is that the claim that one needs to take care of oneself prior to taking care of others, is a reduction of the complexity of the emergence of care. Care, much like attention, is granting signification to that which matters and as such, it is a process of valuation.

Attention does not exist outside of processes of care, much like signification does not exist outside of attention. The value of attention and the care I have for something is that which gives its valuable significance. Much like care, invention needs to be wanted, anticipated and conceived by its agent.[36] Invention in Stiegler is very much influenced by the operative category of the image found in the philosophy of Gilbert Simondon. Invention may generate something else than what might be intended, much like art and science do. But contrary to innovation, invention implies cultivating its pharmacological ambivalence through a technique of care.[37]

For Stiegler, it is within this constant friction and confrontation between innovation and invention, between technogenesis and sociogenesis, that one can find the resources to address the *malaise* of our times, which he defined as being created by the disruptive divorce between computational understanding and reason. Today, to counteract the technological rule of law, one might have to ponder the philosophical question of the dream in the age of disruption, an age defined by the systemic dismantling of the condition of possibility to create collective protentions. Scenes of disruption imply that, in order to critically change our epoch and transform the lack of it into a memorable event, we need to create a distance, the same distance that theory shares with theatre (*theatron*: the place from which actions are seen). In bearing witness to tragic announcements of anthropogenic disruption, one also cultivates a noetic intermittent distance, where the value of bonds (*liens*) can replace capitalist goods (*biens*).

To sustain alternative forms of investment and localities (understood as culture and care), I draw attention to the immensely unpredictable force of dreams: its capability to transduce knowledge into operations of sustainable, collaborative and innovative changes. In order to build localities of becoming, where forms of existence are valued and exceed forms of subsistence. Let me end on a last hypothesis. To form dreams, one has to engage in the organological structure of our times, while developing pharmacological operations that can unleash the power (as in *puissance*) to revolutionize our present condition. This implies that there will be another day, another night and another journey around the sun. It is between the earth and the sun that Bernard concluded his last acting out, his bifurcation.

Post-script

I would like to dedicate this text to a hand-crafted fish I gave to Bernard Stiegler after my dissertation defence in 2016. He hung it above his desk in the moulin d'Épineuil-le-Fleuriel. He named it an exception, an *ange poisson*, because it flies without coming down and called it a fetish because he considered it inhabited by a spirit. In August 2020, that fish must have been petrified by the shared suspension of the standstill, as only death by hanging can achieve such a vertical ataraxy. This image of hanging is the mental supplement I have of an act that occurred prior to its unfolding. The announcement of the suicide preceded Bernard's act. Much like in a tragedy, we saw it coming. Yet, knowing what was going to happen did not prevent such an announcement to be less of a shock. Throughout the years, I had secretly cultivated the wish that for once the Pythia was going to be wrong. And such a wish was my form of care.[38] The image of such a scene is both a projection and its *survivance*, meaning that it is a reminiscence, for *anamnēsis* is both a performative fiction and the creative realm of present and future significations. The mental image I have of the petrified fish – petrified because of its stupor and its incapacity to act, but mainly because of its interrupted intermittence – is an *epochal* scene where the madness of our times exceeded the realms of the *pansable* (curable): 'for *penser*, to think, previously meant *soigner*, to care, to treat.'[39, 40]

Notes

1 On the *pharmakon* as shadow, see Bernard Stiegler, *États de choc: Bêtise et savoir au XXIe siècle*, Paris: Éditions Mille et une nuits, 2012, 58; Bernard Stiegler, *States of Shock: Stupidity and Knowledge in the 21st Century*, trans. Daniel Ross, Cambridge (UK): Polity Press, 32-3.
2 Bernard Stiegler, *Technics and Time, 1: The Fault of Epimetheus*, trans. Richard Beardsworth & George Collins, Stanford: Stanford University Press, 1998, ix.
3 Stiegler, *États de choc*, 28; Stiegler, *States of Shock*, 12.
4 Bernard Stiegler, *Passer à l'acte*, Paris: Galilée, 2003, 42; Bernard Stiegler, 'How I Became a Philosopher', in Bernard Stiegler, *Acting Out*, trans. David Barison, Daniel Ross, and Patrick Crogan, Stanford: Stanford University Press, 2009, 1-35, 19-20.
5 Stiegler, *Passer à l'acte*, 47; Stiegler, *Acting Out*, 21.
6 Stiegler, *Passer à l'acte*, 33; Stiegler, *Acting Out*, 14.
7 Bernard Stiegler, 'Elements of Neganthropology', in Bernard Stiegler, *The Neganthropocene*, ed., trans., and with an introduction by Daniel Ross, London: Open Humanities Press, 2018, 76-91, 79.

8   Anaïs Nony & Bernard Stiegler, 'Bernard Stiegler on Automatic Society, As Told to Anaïs Nony', *The Third Rail Quarterly* 5 (2018): 16-17.
9   Bernard Stiegler, 'What Is Called Caring? Thinking beyond the Anthropocene', in Stiegler, *The Neganthropocene*, 188-270, 201-2.
10 Souleymane Bachir Diagne, *African Art as Philosophy: Senghor, Bergson and the Idea of Negritude*, trans. Chike Jeffers, Calcutta: Seagull Books, 2011, 84.
11 Sara Baranzoni & Paolo Vignola, 'Rhythms of Locality. A Travel through Caribbean Performances and Literature', *La Deleuziana: Online Journal of Philosophy* 10 (2019): 160-77.
12 Bernard Stiegler, *What Makes Life Worth Living: On Pharmacology*, trans. Daniel Ross, Cambridge: Polity, 2013.
13 Bernard Stiegler, *La technique et le temps 1: La faute d'Épiméthée*, Paris: Galilée, 1994, 196 and passim; Stiegler, *Technics and Time, 1*, 188 and passim.
14 Bernard Stiegler, *Pharmacologie du Front National*, suivi du *Vocabulaire d'Ars Industrialis* par Victor Petit, Paris: Flammarion, 2013, 193.
15 Bernard Stiegler, *Uncontrollable Societies of Disaffected Individual. Disbelief and Discredit, Volume 2*, trans. Daniel Ross, Cambridge: Polity, 2013, 6.
16 Etienne Balibar, 'Âme', in *Vocabulaire européen des philosophies*, ed. Barbara Cassin, Paris: Editions du Seuil / Dictionnaire Le Robert, 2004, 80.
17 Anaïs Nony, 'Nootechnics of the Digital', *Parallax* 23/2 (2017): 129-46, 130.
18 Patrick Crogan & Bernard Stiegler, 'Knowledge, Care and Trans-Individuation: An Interview with Bernard Stiegler', *Cultural Politics* 6/2 (2010): 157-70, 165.
19 Bernard Stiegler, *La technique et le temps 2: La désorientation*, Paris: Galilée, 1996.
20 Stiegler, *Pharmacologie du Front National*, 160.
21 Antoinette Rouvroy & Bernard Stiegler, 'The Digital Regime of Truth: From the Algorithmic Governmentality to a New Rule of Law', trans. Anaïs Nony & Benoît Dillet, *La Deleuziana: Online Journal of Philosophy* 3/1 (2016): 6-29.
22 Gilbert Simondon, *L'individuation à la lumière des notions de formes et d'information*, Grenoble: Éditions Jérôme Millon, 2013, 246-7.
23 Bernard Stiegler, 'The Organology of Dreams and Arche-Cinema', in Stiegler, *The Neganthropocene*, 154-71, 156.
24 Don Ihde, 'Visualizing the Invisible. Imaging Technologies', in *Postphenomenology and Technoscience: The Peking University Lectures*, ed. Don Ihde, Albany: State University of New York Press, 2009, 45-62.
25 Jonathan Beller, *The Cinematic Mode of Production: Attention Economy and the Society of the Spectacle*, Lebanon: Dartmouth College Press, 2006, 38.
26 Jonathan Crary, *24/7: Late Capitalism and the End of Sleep*, New York: Verso Books, 2013.
27 Bernard Stiegler, *Dans la disruption: Comment ne pas devenir fou?* suivi d'un *Entretien sur le christianisme* [entre] Alain Jugnon, Jean-Luc Nancy et Bernard Stiegler, Paris: Les Liens qui Libèrent, 2016, 39; Bernard Stiegler, *The Age of Disruption: Technology and Madness in Computational Capitalism*, followed by *A conversation about Christianity with Alain Jugnon, Jean-Luc Nancy and Bernard Stiegler*, trans. Daniel Ross, Cambridge: Polity, 2019, 18.
28 Stiegler, 'Elements of Neganthropology', 80.
29 Personal notes from Stiegler's class on 30 November 2013.
30 Stiegler, *Dans la disruption*, 413; Stiegler, *The Age of Disruption*, 289.
31 Stiegler, 'Elements of Neganthropology', 76.
32 Bernard Stiegler, *Ce qui fait que la vie vaut la peine d'être vécue: De la pharmacologie*, Paris: Flammarion, 2010, 28; Stiegler, *What Makes Life Worth Living*, 12.

33 Nony, 'Nootechnics of the Digital', 129–46, 131.
34 Anaïs Nony, 'Technology of Neo-Colonial *Episteme*', *Philosophy Today* 63/3 (2019): 731–44, 739.
35 Stiegler, *Dans la disruption*, 71; Stiegler, *The Age of Disruption*, 38.
36 Gilbert Simondon, *L'invention dans les techniques: Cours et conférences*, établie et présentée par Jean-Yves Chateau, Paris: Seuil, 2005, 21.
37 Simondon distinguishes the verb *to generate* from the verb *to invent*; the first action belongs to the category of the living (*le vivant*) and the second to the category of the technical object (*l'objet technique*).
38 Martin Heidegger, *Being and Time*, trans. Joan Stambaugh, Albany: State University of New York Press, 2010, 188: 'Ontologically, wishing presupposes care'.
39 Bernard Stiegler, 'What Is Called Caring? Thinking beyond the Anthropocene', trans. Daniel Ross, *Techné: Research in Philosophy and Technology* 21/2–3 (2017): 386–404, 398.
40 The author would like to thank Erika Weiberg, Eliana Vagalau, and the editorial team for their comments and supportive feedback.

Chapter 6

AGAINST SIMPLIFICATION: THE INTERMITTENCE OF LIFE

Gerald Moore

Just prior to his own suicide in November 1995, the last work published by Gilles Deleuze was an all-too-brief excursus, 'Immanence: A Life', on the singularity of vital processes that always escape the confinement of life to the seeming 'transcendence' and 'determination' of individual organs and organisms.[1] At the heart of the essay is an image, taken from Dickens, of beatitude creeping across the lips of a dying wretch, an impersonal, incalculable and excessive force of libidinal possibility that seeps through and triumphs over the apparent entrapment of death. Bernard Stiegler did not quite share Deleuze's commitment to what he saw as the (mere) affirmation of an always-already there vitality of life, understood as revolutionary (but virtual and habitually castrated) desiring-production.[2] He saw both life and desire as fragile and in need of careful cultivation to stave off collapse.[3] He also thought true vitality, at least in the case of exosomatic, 'noetic' animals, occurs only when we use technical objects to shape the world around us and internalize the experience of this work in the form of new knowledge and *savoir-vivre*. The same technologies can simultaneously take hold of, proletarianize and exhaust us, forcing us to adapt to their imperatives without affording us any role in their creation of ourselves and our environments. Technics thus serves as the condition of possibility and impossibility of both desire and *la vie de l'esprit*. Noetic vitality is moreover fleeting at best, glimpsed in exceptions to the general tendency of regression towards entropy, or the dissipation of energy available for work into unharnessable violence and emotional exhaustion. We '*inhuman* beings' become *not-inhuman*, leaping onto Deleuze's plane of consistency '*only intermittently*', which is to say, only when we use our tools to create a quasi-causal agency that wrests us out of unthinking and unproductive habits, before readjusting our behaviours around new habits that will, in time, become equally unthought.[4]

But as we see right up until his final written text, '*Démesure, promesses, compromis*', delivered to Mediapart just before his death in August 2020 and published exactly a month later, Stiegler shares Deleuze's interest in the incalculable – which is to say, in the irreducibility of life to calculations of productivity outputs. He rejects quantificatory attempts to determine the future with such certainty that it effectively disappears, self-defeatingly truncated by the refusal to countenance the unpredictability, the spiralling unintended consequences, of our longer-term Anthropocenic horizons. The obsession with accumulating (big) data to make us visible and facilitate predictive certainty becomes the 'regime of truth' through which 'humanity self-destructs'.[5] This is, on one hand, because of the way that science is rendered complicit in the elimination of intermittence, automating bodies for the purpose of often mindless work, and in so doing transforming noetic life into the manipulable substrate of what Jason W. Moore calls 'Cheap Nature'. The co-optation of science by capitalism reaches its apogee in computational capitalism, on one hand via the threat of replacing us with robots, which pushes us to adapt to long hours of precarious, habitually mindless, production; on the other, through the algorithms that regress us to the automated behaviours of compulsive consumption, while simultaneously coaxing us to surrender what little remains of leisure time to the performance of unpaid labour for the very search engines and social media companies whose services we addictively consume. Like Jason Moore, Stiegler locates the root of this sacrifice of intermittence to expectations of ceaseless productivity in Cartesian rationalism: not just in the ego that declines to recognize its own intermittence, by believing itself permanent and not just coincidental, with the assertion of '*cogito ergo sum*', but moreover in the mindset that treats the *res extensa* of nature as subordinate to the transformations of reason.[6] We can, however, trace the trajectory of this culture of calculation much further back – and, in so doing, ultimately reinforce Stiegler's argument over not only the intermittence of life, but also of a 'primordial diversity' of ways of living that have since been homogenized out of existence.[7] This move will take us back to the start of the Holocene and the first great organological revolution, when the settlement of agriculture made for an unprecedented acceleration in the cumulative human culture that Stiegler conceives in terms of tertiary memory.

For what – to play on the opening question of Deleuze's final essay – is a cultivated field, if not the condition of both the possibility and impossibility of the kind of life of the mind that Bernard Stiegler terms '*noēsis*'? The answer is less straightforward than often assumed. Anthropologists including Jared Diamond have argued that the earliest moves in the history of agriculture and domestication created a basis for social stability and forward planning, which in turn freed up leisure time, enabling a part of the population to devote itself to thought, art and government. Neolithic technologies like blades, baskets, pestles, ovens and waterproof containers for the harvesting, grinding, roasting and storage of seeds, in this respect, are seen to have paved the way for a life of intermittence, spared from the all-encompassing labours of food production – at least for some.[8] The alternative reading, tied to the structural impossibility of permanent *noēsis*, is that Pleistocenic life was already intermittent; that this intermittence was destroyed by the use of agriculture to justify a splitting of populations into a disautomated, privileged class of thinkers who (mis)took their leisure as a permanent condition, and an underclass of exhausted, disease-ridden workers, forced into a life of subsistent servitude. The question faced by elites was the same then as they face now, in the context of a pandemic caused, in no small part, by the exhaustions of a burned-out society, and further beset by a critical loss of confidence in the very science that might just save us. For how long can we stave off the unintended consequences of this separation?

The original intermittent society

> Rather than heralding a new era of easy living, the Agricultural Revolution left farmers with lives generally more difficult and less satisfying than those of foragers. Hunter-gatherers spent their time in more stimulating and varied ways, and were less in danger of starvation and disease. The Agricultural Revolution certainly enlarged the sum total of food at the disposal of humankind, but the extra food did not translate into a better diet or more leisure. Rather, it translated into population explosions and pampered elites, the average farmer worked harder than the average forager, and got a worse diet in return. The Agricultural Revolution was history's biggest fraud.[9]

Perhaps the central contention of Yuval Noah Harari's unexpected bestseller, *Sapiens*, is that human civilization bifurcated for the worse around 10,000 years ago, at the time of the Neolithic agricultural revolution. Harari describes how, prior to the settling of agriculture, our hunter-gatherer ancestors would have spent just a few hours per day in a rather leisurely pursuit of a variety of foraged fruits, seeds and meats whose nutritional value far outstripped what we have endured over the majority of subsequent cultural history. Wheat changed everything, however – as would have come as no surprise to the fastidiously glutophobic Stiegler. Its cultivation demanded vast quantities of land-clearance and back-breaking labour, both human and animal, forcing workers of all species into densely packed 'permanent settlements that would be hotbeds for infectious diseases', the earliest cities.[10] Our ancestors' dependence on a monocultural, carbohydrate, diet moreover led to mineral deficiencies and tooth decay, plus heightened susceptibility to drought and immunological weakness. Echoing Harari, the Yale-based political and agrarian theorist James C. Scott writes that 'the late Neolithic multispecies resettlement camp involved a lot more drudgery than hunting and gathering and was not at all good for your health. Why anyone not impelled by hunger, danger, or coercion would willingly give up hunting and foraging or pastoralism for full-time agriculture is hard to fathom'.[11] Harari's answer to this question suggests time was of the essence. Had the revolution been sudden, then we might have noticed its effects quickly enough to change our minds. But its advent played out via a gradual decrease in infant mortality, the deferral of which only until later childhood coincided with a corresponding growth of anxiety in the face of death. The 'fateful miscalculation' to transition to sedentary agriculture worked as a trap, habituating – we might better say automating – Neolithic peoples into routines of living and expectations of imagined luxury, to which they clung well past the point of their costs outweighing the benefits.[12] Scott's alternative response is less subtle, but more compelling in the 'deep historical' evidence he provides for it. The surrender was not willing, he argues, but the consequence, intended or otherwise, of a creeping culture of simplification that did away with a huge variety of bio-, noo- and technodiversity, *savoir-faire* and *savoir-vivre*, all for the sake of making cultivars sufficiently visible to emerging regimes of taxation. To see how this transformation played out, let us return to what, with hindsight, we might call the original *intermittent society*.

In his seminal 1972 work of speculative anthropology, *Stone-Age Economics*, Marshall Sahlins famously hypothesized that the Palaeolithic era preceding the Agricultural Revolution might well be described as 'The Original Affluent Society'. At least '90% of human history', according to this argument, 'was in no sense a struggle for existence'.[13] Sahlins cites then-contemporaneous studies of Australian bushmen, which showed twentieth-century hunter-gatherers managing to yield a caloric surplus of food despite working only for one to two days per week. The rest of their time was spent 'intermittently' sleeping, resting and doing occasional bits of tool-craft.[14] The absence of more intensive labour would have been enough, he suggests, to generate and legitimate a stereotype of indolence among nineteenth-century colonial settlers, who wondered how the bushmen filled their days before the new arrivals 'taught them to smoke' and left them 'begging' for tobacco.[15] Sahlins raises that last point light-heartedly, but, noting the evidence that wheat, too, is thought addictive – not least in the form of the beer that has historically been used to pay workers – we might still wonder in passing what role addictogens played in the subsequent coercion and automation of labour.[16] A future return to that question will build on the 'beer before bread hypothesis', according to which 'large-scale, likely alcohol-fuelled feasting' and ritual intoxication 'began well before settled agriculture'.[17] More important for now, however, is the re-elaboration of Sahlins's thesis of 'barbarian' affluence by Scott, for whom 'we have surely underestimated the degree of agility and adaptability of our prestate ancestors'.[18]

In speaking of 'agility and adaptability', Scott alludes to the contemporary, highly ideological language of resilience as flexibility, which, as argued by both Bernard and Barbara Stiegler, nowadays tends to consist in a globalized monoculture of market-led solutions: the elimination of intermittence and spare capacity in favour of a narrow range of supposed consistently reliable outcomes, from high-yield crops and just-in-time delivery to a paradoxical fetishization of both nomad mobility and relentless, often deskbound, labour.[19] At the heart of this ideology of adaptation is an emphasis on the individual's capacity to absorb environmental perturbation, or what Georges Canguilhem, defining health, called the 'margin of tolerance for the inconstancies of the milieu'.[20] And yet the discourse ultimately neglects the very means by which we become adaptable, namely the ability to transform environments that Canguilhem terms normativity, identified as the fundamental activity of life itself: 'life, as not only subject to the milieu but also as institution of its own milieu'.[21] The capacity of the organism to organize its surroundings into a milieu, or lifeworld, through the 'institution of new norms' means that it is not simply passively adapting to environments imposed from without. In the case of humans, our normativity is specifically technical, playing out through our use of technology and the acquisition of transformative, anti-entropic knowledge that we gain from mastering tools.²²

Expanding on Canguilhem, Stiegler argues that the institution of norms through technics simply goes by the name of 'work', understood as transformative of both the 'who' and the 'what', and theorized in opposition to the entropic labour of adaptation better described as 'employment'. 'Perhaps we can call "work" everything that, in the technical form of life, which is also its noetic form, differs from and defers entropy by intensifying "neganthropy".'²³ To work, in other words, means to translate the potential (*dynamis*) opened up to thought by our tool-use into the building of a future, towards which we can project ourselves in the form of desire. Elsewhere, this deferral of entropy translates into a relaxation of Darwin's natural selection, by allowing the modified environment to reduce or take the strain of the selection pressures acting on the organism.²⁴ By contrast, strip down the possibilities of self- and world-transformation, and we experience a dramatic narrowing in our margins of tolerance for change.

To make the same point, Scott draws on the contemporary evolutionary theory of 'niche construction', which emphasizes both the inadequacy of adaptation for understanding the participatory role organisms play in ecosystem engineering,²⁵ and the way niche-constructing organisms can decrease local entropy levels within their milieus, for example by enhancing food-production, albeit at the cost of increasing entropy beyond them.²⁶ Reworking Sahlins, what we might call the prestatal intermittent society revolves around 'a deliberate disturbance ecology in which hominids create, over time, a mosaic of biodiversity' that lifts them above mere subsistence.²⁷ Far from adhesion to a single way of living, be that the allegedly precarious existence of 'just' hunting and gathering, or the surplus-generating, hence supposedly failsafe, security of domestication and sedentary agriculture, Neolithic life consisted in a hybrid multiplicity of strategies for flourishing. In practice, so-called hunter-gatherers of the Fertile Crescent were more like hunter-gatherer-forager-pastoralist-fishers and landscape engineers, engaged in everything from flood-retreat and slash-and-burn agriculture to the domestication of cattle and wild grains. Much is made of the technologies that rendered sedentarism possible, and whose accumulation by nomads, hitherto unable to afford the weight and baggage of anything but minimal equipment, was in turn made possible by sedentarism: sickles and handled blades for harvesting, baskets for carrying, pestles for grinding and underground, often waterproof, pits for storage, among others.²⁸ But their invention, alongside techniques for building 'drive corridors', 'weirs, nets and tools for smoking, drying and salting', largely preceded the dominance of domesticated agricultural grain states.²⁹ We can accordingly see that the 'intermittence' of Neolithic intermittent society offers more to a Stieglerian reading than linguistic happenstance.

First developed in around 2003–4, in *Acting Out* and the closing chapters of *The Decadence of Industrial Democracies*, Stiegler's theory of intermittence builds on a reworking of the Aristotelian hierarchy of three distinct, but overlapping vegetative, sensitive and intellective (or 'noetic') souls.³⁰ At the heart of this theory is the idea that only the divine 'unmoved mover' is permanently noetic. The majority of animals spend most of their time sleeping and eating, which is to say, vegetating, becoming intermittently reactive to sense impressions when looking to reproduce, Aristotle suggests. We, supposedly higher, mortals alternate between vegetation and engagement in acts of experiencing the world and are, for the most part, only *potentially* noetic, except for fleeting moments of elevation when, through the translation of intellectual potential (*dynamis*) into the energetic output (*energeia*) of work, we participate in the divine.[31] Stiegler insists that the three souls are compositional, not oppositional: knowledge emerges in the abstraction from sensory experience, and can be destroyed when our sensations of the world are prescribed by technologies that no longer afford us the prospect of interpreting how we use them. The intermittent dormancy that to the eye of the colonial explorer looks like the vegetative subsistence of lazy hunter-gatherers, cannot be separated out from their sensorial reading of flora, fauna and landscapes, which paves the way, in turn, for the sublimation of sensory experience into the transformative action of niche- and self-construction, mediated by the mastery of myriad Neolithic tools. With the reorganization of societies around sedentary farming, however, this diverse range of technical skills gave way to a life devoted exclusively to the 'harrowing and sowing' of fixed field farming. Scott describes 'the late Neolithic revolution [...] as something of a deskilling', where the know-how needed to harvest 'a wide spectrum of wild flora' and fauna is sacrificed for the narrower specialisms put to work in cultivating 'a handful of cereals and a [...] handful of livestock'.[32] Their loss, in other words, coincided with proletarianization and a dramatic collapse in the (bio-, noo- and techno-diverse) modes of resilience provided by alternative routes to sustenance. As we shall see, it also coincided with the sacrifice of Neolithic peoples' ability to 'read', or interpret, and thereby transform their milieus to their forced adaptation to a new environment, imposed on them from above for the purpose of making the people themselves legible.

Against calculation – From intermittence to Covid-19

Far from being a uniquely modern, capitalist, phenomenon, it was calculation that spelled the disappearance of the original intermittent society. Tubers can be planted in secret and left in the ground until needed; legumes grow and ripen continuously, making it hard to calculate yield throughout the year. 'Looked at from the perspective of a state tax collector', 'hunting and gathering, maritime fishing and collecting, horticulture, shifting cultivation, and specialized pastoralism' were all 'fiscally sterile', or 'what might be called nonappropriable subsistence activities', whose products 'were so dispersed and mobile, and their "takings" so diverse and perishable, that tracking them, let alone taxing them, was well-nigh impossible'.[33] Even if offering worse nutritional value for more labour-intensive cultivation than lentils, chickpeas and potatoes, among others, cereals like wheat, barley, millet and maize grew above ground in the same fields, year-on-year, in seasonably predictable, regular, yields, which could moreover be readily divided into precisely measurable subunits of husks

6. Against Simplification

and grains. They were more 'legible', in Scott's memorable phrase, which is to say, 'visible, divisible, assessable, storable, transportable, and "rationable"'.[34] This legibility is what secured the widespread imposition of settled agriculture, in spite of the huge costs of mandating their cultivation: the slave labour needed to produce a surplus, the ensuing crowding and disease that become inseparable from 'multispecies resettlement camps' and the loss of both resilience and 'the exuberant diversity of livelihoods' that coerced monoculture entailed.[35] Those doing the mandating, we might add, were part of a new and leisured governing class, afforded the time to think, create and pursue specializations in craft, writing and science. Intermittence became the privilege of a select few, no longer spread throughout the population with rough equality. The resulting hierarchies could endure even during agricultural off-seasons, when the feudal lords of intermittent agricultural workers coercively redeployed them to Sisyphean building projects, like the first city, Çatalhöyük and later the pyramids, for which their exhaustions were paid in meat and beer.[36] Such projects undoubtedly owed much to farming. The evolutionary geographer Jared Diamond has described how the food surpluses of settled agriculture allowed for the emergence of 'complex social organization' and 'complex centralized societies [...] uniquely capable of organizing public works' like the pyramids.[37] Returning to Scott, however, we might wonder how far this really amounts to a complexification, as opposed to a lethal simplification of social organization.

Against the Grain's focus on states' 'making legible' in the Neolithic Fertile Crescent is borrowed from and extends the thesis originally laid out in James Scott's seminal work of anarchist theory, *Seeing Like a State* (1998), on the simplificatory schemas imposed by state apparatuses for the sake of facilitating government. Scott's earlier book begins with modernity and the concretization of society around science and capitalism. He focuses, in particular, on the eighteenth-century emergence of commercial forestry and the ultimately disastrous strategies of artificial selection that, in the light of his later work, we can read as a continuation of the practices employed at the time of the Neolithic revolution. With the passage of time, the promotion of a small range of visibly consistent, albeit fragile and high-maintenance crops over the engineered mess of biodiversity that had preceded them, is expanded and refined to catastrophic perfection. When the demand for wood as fuel began to exceed supply in Northern Europe, nascent practitioners of forestry science responded by replacing old-growth forests, made up of a vast range of different trees, with standardized and uniformly regimented, fast-growing and high-yield softwood monocrops, like Norwegian spruce and Scotch pine, planted in neatly visible, countable rows. The forestry equivalent of fixed fields was moreover cleared of all undergrowth and fallen leaves to facilitate access, harvesting and replacement.

> In the short run, this experiment in the radical simplification of the forest to a single commodity was a resounding success [...] The productivity of the new forests reversed the decline of the domestic wood supply, provided more uniform stands and more usable wood fibre, raised the economic return of forest land, and appreciably shortened rotation times.[38]

Things nonetheless went badly wrong – almost immediately for the localities hitherto dependent on the woodland, though it would take the eighty-year duration of tree rotation for the economics to unravel. 'The monocropped forest was a disaster for peasants, who were now deprived of all the grazing, food, raw materials, and medicines that the earlier forest had afforded',[39] who were proletarianized, in other words, by the destruction of the locality to which their knowledge strictly pertained. And while the first generation of trees far exceeded expectations of growth, this is now thought to be because of the fertile topsoil produced by the old-growth leaf mulch, which housed a vast array of biomass that disappeared once the economic imperatives of forestry management decreed the removal of deadwood and ground cover. The depletion of tree diversity also made for decreasing resilience, with mono-arboriculture revealing itself particularly susceptible to stressors like disease and fluctuations in the weather. As epidemic proportions of species-specialized pests, plus drought, cold and soil-depletion, took hold, what had begun as processes of simplification soon became entangled in increasingly more complicated attempts to manage the unintended consequences of forestry science. Ever greater quantities of 'fertilizers, insecticides, fungicides, or rodenticides' failed to offset declining rates of growth, leading to the artificial raising and reintroduction of previously resident organisms, naively expected to thrive in 'impoverished habitats', or locations stripped of their once defining locality.[40] 'In this case, "restoration forestry" attempted with mixed results to create a *virtual* ecology, while denying its chief sustaining condition: diversity'.[41]

Scott's collapse in diversity principally refers, here, to the loss of the resilience provided by a forest's biodiversity, but another, more Stieglerian, take on the phrase is also implicit – notably, in the form of the monoculture of intellectual and technical practices that characterizes the industrialization of forestry. This is the focus of Stiegler's reconstruction of much the same argument, in *Constituer l'Europe 2* from 2005, where he lays the groundwork for a concept of 'noodiversity' that captures the productive variety of irreducibly local, ecologically specific, forms of technical practice and the thought to which they give rise, to which he subsequently returns only towards the end of his life.[42] Without making Scott's link of continuity between the industrial and Neolithic revolutions, Stiegler dates the nascence of a cultural obsession with calculation to somewhere around 1780, coinciding – as per Scott – with a growing interest in management and the technoscientific adaptation of life to the exigencies of profit. The shift, he suggests, is signalled by the rise of the factory and an ensuing movement away from the 'performativity' of work, construed in terms of niche- and self-construction, and towards the modelling of labour around calculable performance. The horse and later the machine become the motors 'to whose performance the proletarian will have to adapt their own performance',[43] adjusting to function mechanically and automatically, which is also to say, unthinkingly – both because they have no time to think while labouring, and because exhaustion leaves them unable to do so. Forced to surrender both the intermittence of rest and, accordingly, the prospect of *noēsis* in order to compete with automata in a manufactured struggle for survival, the worker is 'bestialized', which is to say, reduced to a combination of imprinted habits and sensory responses.⁴⁴ This 'homogenization' of worker behaviours around a model of automated, competitive, performance is 'entropic'. 'Negentropy', by contrast, 'is the diversification of types, just as indispensable to social life as biodiversity is in increasing the vitality of organisms.'⁴⁵

Fast forward another 200 years and we see the impact of man-made *anthropie* on a society collapsing under the weight of its exhausted monoculture of resilience. The current pandemic of Covid-19 is already prefigured in the contagion and exhaustion that result from the calculations of the agricultural and industrial revolutions. The exploited labour of the early states goes hand-in-hand with the proliferation of contagion. The crisis of managed forestry similarly foreshadows the risk of diseases – including zoonotic ones unearthed by deforestation – ripping through genetically homogenized megafarms, where weakened poultry and livestock are so heavily dosed up with prophylactic antibiotics that pathogens have become resistant.⁴⁶ Megafarm transmission may have played a comparatively minor role in the case of Covid-19, but the prevalence of comparable conditions in urbanized human societies has placed us in a similar position of precarity. We have known since the early days of Coronavirus just how far we are dealing not simply with a 'pandemic', but moreover a 'syndemic', which is to say, a contagion whose virality results from social circumstance just as much as biology.⁴⁷

One's chance of being affected by SARS-Cov-2 is subject to a range of markers including housing size and density, access to greenspace and one's type of work. It is also inseparable from our obsessive, (self-)exploitative,⁴⁸ availability for labour: the cytokine storms thought linked to Covid's fatality occur most vehemently in those with poor mental health,⁴⁹ sedentary lifestyles,⁵⁰ sleep deprivation, burnout⁵¹ and obesity-inducing diet, including diabetes⁵² – all of which symptomatize, in turn, the deleterious organization of Western societies around a concept of life defined by non-stop productivity. We needn't look far to see an analogy of the pine tree, propped up by pesticide and fertilizer as a condition of meeting its required yields of wood fibre, in the figure of the underslept and overstressed worker, denied the repose of intermittence and kept running not just by the fear of losing their deliberately precarious employment, but by a panoply of addictogenic *pharmaka* that serve to expand our margins of tolerance for environmental perturbation: coffee, sugar, screen time, painkillers and antidepressants, among others. The syndemic has rammed home the message that we already knew but preferred not to see play out in milieus of enforced adaptation right across the planet, namely that being 'always-on' is the cause of widespread breakdown. Following several global lockdowns, what now stares us in the face is the forced choice of a return to life as intermittence, or an only intermittently functioning society.

Conclusion: Against simplification

The theoretical biophysicist Stuart Kaufman has argued that the emergence of life should no longer be deemed miraculously improbable. 'Under rather general conditions, as the diversity of molecular species in a reaction system increases, a phase transition is crossed beyond which the formation of collectively autocatalytic sets of molecules suddenly becomes almost inevitable. If so, we are birthed of molecular diversity.'[53] For Bernard Stiegler, 'in the era of the hegemony of probability calculations', which is to say, of neoliberal monoculture and globalized homogeneity, the diversity necessary for life is itself what has become improbable. 'The improbable is diversity – in this case, biodiversity, noodiversity. The probable', played out *ad infinitum* through the same sets of broken market solutions, 'is the entropic tendency towards the elimination of the diverse.'[54] Our overdependence on a single lifestyle has led to critical exhaustion – and with it, the ever-greater probability of a dramatically improbable event escaping attempts to render life both fully predictable and automatic.

It is the fear of this kind of unintended consequence that leads someone like Frédéric Neyrat to baulk at the idea of nature as a 'construction', desanctified to the point of eliminating the psychological barriers to geo-engineering technologies like atmospheric phosphate-seeding, which offer 'solutionist' quick fixes for climate change, but at the cost of potentially uncontrollable repercussions.[55] The combination of Scott and Stiegler allows us to posit an alternative to this rather simplistic conception of nature, however. Speaking in an interview in 2016, Stiegler rejected the idea of 'nature' as somehow distinct from artificial selections, which is to say, from the tools through which we categorize, understand and manipulate the organic world. 'The problem is not "nature and culture". It is the processes of individuation': of the chemical diversity that gives rise to new organs and organisms when their interactions pass the threshold from an aggregate of elements to a new functional whole; of the mutual constitution of the *who* and the *what*, whose combination gives rise to the reinvention of both, and so on.[56] We see this already in the diversity of Paleolithic food strategies, where the multiplicity, or noodiversity, of styles of hunting and farming contributes to the enhancement of biodiversity. Complications come not from artificial selection *per se*, either in the narrow, Darwinian, sense of selective breeding, or the expanded Stieglerian one of technics as 'the pursuit of life by means other than life'.[57] Rather, they arise when the effect of these selections is simplification.

We need not see nature as either metaphysically (or biologically) distinct, nor even as *simply* constructed. We should see it, rather, as a vast and diverse accumulated reservoir of organic retentions and inadvertent protentions, a memory for the future with enough iterations behind it to have generated anticipatory solutions that can only be dreamed of by our own efforts to manage the consequences of its depletion through simplification. The history of that simplification is inseparable from overcompensatory attempts to rein in an increasingly unmanageable array of unintended consequences, all seemingly converging towards a near-future of monumental collapse. None of which is to suggest that we should or could – except, perhaps, in the aftermath of looming ecological disaster – lose ourselves in a fantasy of Neo – or even Paleolithic living. The challenge is rather to think through the role that contemporary technologies, and above all, a new, noodiverse society of intermittence, might play in counteracting that history.

Notes

1. Gilles Deleuze, 'Immanence: A Life', in Gilles Deleuze, *Pure Immanence: Essays on a Life*, trans. Anne Boyman, New York: Zone, 2003, 28–31.
2. On this Deleuzo-Guattarian formulation of desire, see, for example, Gilles Deleuze & Félix Guattari, *Anti-Oedipus: Capitalism and Schizophrenia*, trans. Robert Hurley, Mark Seem & Helen R. Lane, London: Continuum, 2004, 126–7.
3. On Stiegler's habitually implicit critique of Deleuze (and Guattari), see Gerald Moore, 'Adapt and Smile or Die!', in Christina Howells & Gerald Moore (eds.), *Stiegler and Technics*, Edinburgh: Edinburgh University Press, 2013, 17–33, 25–6.
4. Bernard Stiegler, *Taking Care of Youth and the Generations*, trans. Stephen Barker, Stanford: Stanford University Press, 2010, 168–70.
5. Bernard Stiegler, 'Démesure, promesses, compromis 1. Crédit et certitude', *Mediapart* (5 September 2020), available at: https://blogs.mediapart.fr/edition/les-invites-de-mediapart/article/050920/demesure-promesses-compromis-13-par-bernard-stiegler.
6. On the Cartesian roots of 'Cheap Nature', see Jason W. Moore, 'The Capitalocene, Part I: On the nature and origins of our ecological crisis', *The Journal of Peasant Studies* 44/3 (2017): 594–630, 600–5, available at: https://www.tandfonline.com/doi/abs/10.1080/03066150.2016.1235036?journalCode=fjps20.
7. Bernard Stiegler, 'Démesure, promesses, compromis 2. Incertitude et indétermination', *Mediapart* (September 2020), available at: https://blogs.mediapart.fr/edition/les-invites-de-mediapart/article/070920/demesure-promesses-compromis-23-par-bernard-stiegler.
8. Jared Diamond, *Guns, Germs and Steel: A Short History of Everybody for the Last 13,000 Years*, London: Vintage, 2017, 116–17.
9. Yuval Noah Harari, *Sapiens: A Brief History of Humankind*, London: Harvill Sacker, 2014, 79.
10. Harari, *Sapiens*, 87.
11. James C. Scott, *Against the Grain: A Deep History of the Earliest States*, New Haven: Yale University Press, 2017, 18.
12. Harari, *Sapiens*, 86.
13. David Graeber, 'Foreword', in *Stone Age Economics*, ed. Marshall Sahlins, London: Routledge, 2017, xi.
14. Sahlins, *Stone Age Economics*, 19–20.
15. E.M. Curr, cited in Sahlins, *Stone Age Economics*, 23.
16. Paola Bressan & Peter Kramer, 'Bread and other edible agents of mental disease', *Frontiers in Human Neuroscience* 10/130 (2016), available at: https://www.ncbi.nlm.nih.gov/pmc/articles/PMC4809873/.
17. Edward Slingerland, *Drunk: How We Sipped, Danced and Stumbled Our Way to Civilization*, New York: Little, Brown Spark, 2021, 108.
18. Scott, *Against the Grain*, 61.
19. On the discourse of adaptation in Stiegler, see Moore, 'Adapt and Smile or Die!'; see also Barbara Stiegler, '*Il faut s'adapter!*': *Sur un nouvel impératif politique*, Paris: Gallimard, 2019.
20. Georges Canguilhem, *The Normal and the Pathological*, trans. Carolyn R. Fawcett & Robert S. Cohen, New York: Zone, 1991, 199 (translation modified).
21. Canguilhem, *The Normal and the Pathological*, 227 (translation modified).
22. Canguilhem, *The Normal and the Pathological*, 177.

23 Bernard Stiegler, *Automatic Society, Volume 1: The Future of Work*, trans. Daniel Ross, Cambridge: Polity, 2016, 193.
24 Bernard Stiegler, 'L'être soigneux', in Jean-Paul Demoule & Bernard Stiegler (eds.), *L'avenir du passé*, Paris: La Découverte, 2008, 15–25, 22.
25 Richard Levins & Richard Lewontin, *The Dialectical Biologist*, Cambridge: Harvard University Press, 1985, 97–106.
26 F. John Odling-Smee, Kevin N. Laland & Marcus W. Feldman, *Niche Construction: A Neglected Process in Evolution*, Princeton: Princeton University Press, 2003, 190.
27 Scott, *Against the Grain*, 40.
28 Diamond, *Guns, Germs and Steel*, 116–17.
29 Scott, *Against the Grain*, 65–6.
30 Bernard Stiegler, *The Decadence of Industrial Democracies. Discredit and Disbelief, Volume 1*, trans. Daniel Ross & Suzanne Arnold, Cambridge: Polity, 2011, 133–6; see also Aristotle, *De Anima, Book II* (all editions).
31 Aristotle, *De Anima, Book III* (all editions); Bernard Stiegler, 'How I Became a Philosopher', in Bernard Stiegler, *Acting Out*, trans. David Barison, Daniel Ross, and Patrick Crogan, Stanford: Stanford University Press, 2009, 1–35, 12–14; Stiegler, *The Decadence of Industrial Democracies*, 132–5.
32 Scott, *Against the Grain*, 92.
33 Scott, *Against the Grain*, 135.
34 Scott, *Against the Grain*, 129–30.
35 Scott, *Against the Grain*, 127.
36 Alison George, 'The world's oldest paycheck was cashed in beer', *The New Scientist* 3079 (25 June 2016), available at: https://www.newscientist.com/article/2094658-the-worlds-oldest-paycheck-was-cashed-in-beer/; also Jonathan Shaw, 'Who built the pyramids?', *Harvard Magazine* (July-August 2003), available at: https://www.harvardmagazine.com/2003/07/who-built-the-pyramids-html.
37 Diamond, *Guns, Germs and Steel*, 311–12.
38 James C. Scott, *Seeing Like a State: How Certain Schemes to Improve the Human Condition Have Failed*, New Haven: Yale University Press, 2020, 19.
39 Scott, *Seeing Like a State*, 19.
40 Scott, *Seeing Like a State*, 20.
41 Scott, *Seeing Like a State*, 21.
42 See, for example, Bernard Stiegler, 'Elements of Neganthropology', in Bernard Stiegler, *The Neganthropocene*, ed., trans., and with an introduction by Daniel Ross, London: Open Humanities Press, 2018, 76–91, 77–81.
43 Bernard Stiegler, *Constituer l'Europe 2: Le Motif européen*, Paris: Gallimard, 2005, 51.
44 Stiegler, *Constituer l'Europe 2*, 51.
45 Stiegler, *Constituer l'Europe 2*, 64.
46 See, for example, Rob Wallace, *Big Farms Make Big Flu*, New York: Monthly Review Press, 2016.
47 See Barbara Stiegler, *La Démocratie en pandémie*, Paris: Gallimard, 2021, 1–7; Richard Horton, 'Covid-19 is not a pandemic', *The Lancet* 396/10255 (6 September – 2 October 2020), 874, DOI available at: https://doi.org/10.1016/S0140-6736(20)32000-6.
48 Byung-chul Han, 'The Tiredness Virus', *The Nation* (12 April 2021), available at: https://www.thenation.com/article/society/pandemic-burnout-society/.
49 Maxime Taquet, Sierra Luciano, John R. Geddes & Paul J. Harrison, 'Bidirectional associations between COVID-19 and psychiatric disorder: retrospective cohort

studies of 62 354 COVID-19 cases in the USA', *The Lancet Psychiatry* 8/2 (February 2021): 130–40, DOI available at: https://doi.org/10.1016/S2215-0366(20)30462-4.

50 Robert Sallis, Deborah R. Young, Sara Y. Tartof, et al., 'Physical inactivity is associated with a higher risk for severe COVID-19 outcomes: a study in 48 440 adult patients', *British Journal of Sports Medicine* (April 2021), DOI available at: https://doi.org/10.1136/bjsports-2021-104080.

51 Hyunju Kim, Sheila Hegde, Christine LaFiura, et al., 'COVID-19 illness in relation to sleep and burnout', *BMJ Nutrition, Prevention & Health* (March 2021), DOI available at: https://doi.org/10.1136/bmjnph-2021-000228.

52 Livio Luzi & Maria Grazia Radaelli, 'Influenza and obesity: its odd relationship and the lessons for COVID-19 pandemic', *Acta Diabetologica* 57/6 (5 June 2020): 759–64, DOI available at: https://doi.org/10.1007/s00592-020-01522-8.

53 Stuart A. Kauffman, *Investigations*, Oxford: Oxford University Press, 2000, 35.

54 Bernard Stiegler, 'Noodiversity, Technodiversity', trans. Daniel Ross, *Angelaki* 25/4 (2020): 67–80, 71, DOI available at: https://doi.org/10.1080/0969725X.2020.1790836.

55 Frédéric Neyrat, 'Critique du géo-constructivisme', *Multitudes* 56 (2014): 37–47, available at: https://www.cairn.info/revue-multitudes-2014-2-page-37.htm.

56 Bernard Stiegler, 'Artificial Selection and the Function of Science', in *Learning to Live Again*, ed. Bernard Stiegler, Martin Crowley, Ian James & Gerald Moore, Cambridge: Polity, forthcoming 2022.

57 Stiegler, *Automatic Society, Vol. 1*, 13–14; see also Bernard Stiegler, *Technics and Time, 1: The Fault of Epimetheus*, trans. Richard Beardsworth & George Collins, Stanford: Stanford University Press, 1998, 135–7.

Chapter 7

STIEGLER'S HANDS: TERTIARY RETENTIONS AND THE BELIEF OF REASON

Paul Willemarck

Recalling Bernard

If we wish to think the death of Bernard Stiegler using his concept of mnemotechnics (tertiary retention) we are obliged to say that his passing makes no difference. Tertiary retentions remain, but Bernard Stiegler is dead, and we know this makes an immense difference. We feel the passing of our friend as a privation that no trace can heal. We can mourn, we can commemorate, but his *persona*, shall we say, his *psychē*, his soul, is not there anymore for us to attend to. To approach this grief, we need to:

1. understand that what he did was more than exappropriating the classical concept of technics in a differential logic;
2. examine what tertiary retentions are and what they are unable to sustain; since tertiary retention is the very concept of something that is kept, the experience of our mourning can probably help us in understanding what it leaves out;
3. enquire how the experience of his death may point beyond the legacy of his concept of tertiary retention to another contribution of his thought which is his interpretation of mortality as a mode of being that can be historial because it is historical; the expression that famously indicates this twist in the understanding of our being in the world is 'the necessary default' ('*le défaut qu'il faut*').

Stiegler proposed the concept of tertiary retention (or tertiary *souvenir* as he also called it) as a reversion of Husserl's position on the technical support of scientific knowledge into his earlier analysis of time-consciousness, inflected through the analysis of tool-being in Heidegger's *Being and time*. This allowed him to renew Derrida's insight into the necessity of a trace. Perception occurs in time, and as such it is always already lost as time goes by. Only a trace as Derrida would have it, or a technical support as Stiegler would say, can allow us to fixate an experience, rendering it accessible in another moment. But the technical trace of an experience *is* not *this* experience, it is only a material support that allows us to recover a similar, though not an identical, experience, the original experience being forever lost.

Although Stiegler never published the text where he wrote thematically about his concept of necessary default, its idea is introduced from the very beginning of his work. The text was called *Idiocy*, and was intended as the 7th volume of *Technics and Time*.[1] There are two features in the history of this text that might throw light on its subject. First, as Stiegler writes in the introduction to the 2nd edition, *Idiocy* was to be the seminal text of *Technics and Time*. Second, the reason why time and again the thematic development of the notion of the necessary default has been purported is said to be a *lack of evidence*. This throws a different light on *Technics and Time* since it implies that it would not be tertiary retention that is the central notion of the project of *Technics and Time* but necessary default. Indeed, perhaps the reason the evidence for this default is missing lies in the very form of this idiocy, that is, it being a lack, a privation, making its absence necessary.

What is this lack of evidence, this privation? In the text of *Technics and Time* it is obviously a lack of conceptual development or explanation on the theme of idiocy. The mythological account, the story of Epimetheus's fault, could not suffice to support the conceptual realm of the statements in *Technics and Time*.[2] But apart from the textual default, there must be, attached to the very fact of the idiocy, a default that affects life itself. Here is the hypothesis that we put forward: the necessary default of idiocy is a privation of something that could not anymore be provided for. In fact, this hypothesis is just a reformulation of Stiegler's thesis on the fault of Epimetheus causing that of Prometheus and eventually the default of humanity. But then, we have to be attentive to the form of the necessary default in this formulation. The notion 'necessary default' inverses Hegel's paradigm of privation in which the missing part is always promised as a potential that has to be realized or actualized. In the story of Epimetheus, privation does not contain potentially the tools that will factually condition the becoming of humanity, and when they will be provided (by Prometheus) privation will return, but will be determined by the character of the tools as much as by the privation that necessitated them. As in Hegel, the default will get history on its way, but unlike in Hegel the default will not provide for what is missing. Instead of conceiving the default dialectically, Stiegler reformulates it as a repetition in a history of technics. It is a contingent process that has as much to do with chance or accident as it has with its conditions of possibility and the necessities of life. Technics will produce the promise of an empowerment, but eventually it cannot undo the human condition, mortality. On the contrary, while empowering these mortal beings with possibilities that reach far beyond the limits of their natural condition, technics heightens their awareness of death and loss. This is a decisive step of Stiegler in the intensification of the logic of default, in its extension from the logic of trace (as in Derrida) to that of all technics.

Equipping Aristotle's soul

Stiegler thought we can only access our own life if we take into account the orthothetics of the historical position we inherit. He understood this requirement as the historical task to live up to one's idiotext. The idiotext is everything one is missing in the necromass one inherits: the idiosyncrasy or accidentality of one's situation. The accidentality of one's situation is always contingent in view of the orthothetics one is inheriting. It is one's necessary default, it is never a given, it is always to come and, in this sense, the *orthothesis* of our idiosyncrasy is always a protention of what is missing and what needs to be done.

In a sense, one could read *Technics and Time* in two ways: one that would stress the role of technics and another that would stress the role of time. Because it is so pervasive, having read the book, one remains mostly with the story of the maddening evolution of modern technology and its consequences, but just as pervasive in this book is the story of time, which is the story of the *idios* that does not succeed to live up to his text, his idiotext. And it is not surprising then that it is precisely this part we are missing with Stiegler's passing, since the *idios* is what is lacking evidence again and again. But what is more: it even applies to ourselves and to our thoughts and knowledge about ourselves. If there is a lesson of what Stiegler thought, it is surely that we don't experience most of what we think. We perceive in a dynamic of loss, and time-consciousness necessitates a technical support; the task of knowledge transforms into a problem of access.

Stiegler conceived of the idiotext in prison,[3] long before *Technics and Time* was written. The first major development we know of came in the fourth part of his doctoral thesis *Epimetheus fault*, entitled 'The Idiot'. In the three published volumes of this project, this part is still missing (it was to be, as mentioned, the seventh volume). Stiegler published two very similar thematic texts on the concept of idiotext: the first in *Césure* in 1995,[4] and the second in *Intellectica* in 2010.[5] '*Idios*' in the term idiotext signifies singularity.[6] What interests us here, is Stiegler's analysis in 'The Idiot' of Aristotle's theory of the soul. We choose to analyse two paragraphs (Chap. 4, §§15–16) of the fourth part of his doctoral dissertation entitled 'The idiot'. His interpretation of *De Anima* is heterodoxical but important to understand the dynamics of deconstruction in his work. In rearticulating Aristotle's theory of the layering of the soul, Stiegler shows a potentiality of life that will not be preserved in tertiary retentions.

In his analysis, Stiegler examines the movement of the souls in respect to the impassible (§15), and the function of the *idios* with respect to the community of the senses (§16). Stiegler starts by saying that he identifies the soul in De Anima with what he calls idiotext.[7] If it does nevertheless also concern the sensitive and vegetative soul, it is because there cannot be a noetic soul without these. In a sense vegetative, sensitive and noetic soul are cumulative layers, with at its base the vegetative soul being shared by all living beings. According to Stiegler, Aristotle asks *where* the movement of the soul takes place; the answer being: in its medium (milieu). Stiegler proceeds affirming that the relation of the soul to the place of its movement is one of assimilation.[8] The vegetative assimilates its medium (milieu) where it is, the sensitive assimilates it through local movement and the noetic by producing it (*l'instruisant*). That all living beings have to assimilate the medium (milieu) in which they move is the heterodoxical originality of Stiegler's approach, the most interesting part of his analysis, but also the expression of a materialism he found difficult to assume. It is inspired by Derrida's seminar on the *Timaeus* which would be published as a book entitled

Khôra,[9] shortly after Stiegler presented his dissertation in 1993. Derrida's reading showcases the difficulty of understanding the nature of the receptacle where the ideas come into the world if this receptacle is to be of no influence on the character of these ideas. It is the cosmogonic formulation of the paradox, already identified by Anaxagoras, of the medium of perception that should not interfere with the nature of what is being perceived. Employing Aristotle's layering of the soul, Stiegler reformulates this paradox as the problem of the self-understanding of the noetic soul (i.e. as idiotext).

The text starts with the affirmation that the subject of *De Anima* is the relation of each type of soul with its environment (milieu). The environment Stiegler invokes is the associated milieu of Canguilhem and the use of it by Simondon articulating his concept of individuation, but for Stiegler it is also associated to Hegel's interpretation of mediation (medium). Aristotle's theory says that, if they are to live, all living beings need to nourish themselves, which they do by assimilating their milieu, as plants do.[10] The specificity of this operation is that nourishment assimilates the form of other beings together with their matter. The sensitive soul of living beings assimilates its milieu through impression. Perception is characterized by the fact that the senses are sensible to the form of things without the corresponding matter. The noetic soul of living beings assimilates itself to the immutable forms it actually thinks. But access to these forms is barred and this is why Stiegler will say that 'the medium is what should be taken in consideration when the soul is in its noetic reflection.'[11] According to Stiegler, the noetic sense is not given, it has to be produced in an articulation between perception, re-memorization and traces of former experiences, resulting from forgotten perceptions. And since the noetic unity of thought has to be produced, it will be the unity of an aim, the desire of a unity. Stiegler reformulates this as follows: 'default AS eidos and EIDOS as default'.[12] The production of the noetic is animated by the ideal of the unity of thought that is an impassibility, but the unity of thought has to be produced time and again.

In *De Anima*, the retentional character of the noetic soul is only noted in passing, since Aristotle tries to analyse the proper of each soul in it, and the retentional character of the noetic soul has to do with the mortal, that is, mixed nature of humans. But there is at least one other Aristotelian text that gives a clear account of the retentional character of the noetic soul and that is the last chapter of book 2 of *Analytica Posteriora*. Enquiring into the ground of all scientific knowledge (*epistēmē*) Aristotle queries the origin of the immediate premises out of which demonstrative knowledge can emerge. Were these premises innate, it would imply 'that we possess apprehensions more accurate than demonstration and fail to notice them'.[13] This not being the case, knowledge must be produced by induction.[14] The immediate premise out of which we develop all skills (*technai*) and knowledge (*epistēmē*) is identified in sense-perception. We develop a single experience out of memory of sense-perceptions.[15] This single experience is the earliest universal, a single identity made out of many perceptions. Aristotle distinguishes between different ways of grasping these universals amongst which some are subject to error, like *doxa*. Only two escape this flaw: *epistēmē* and *nous*. And since *epistēmē* cannot grasp immediate premises, it is only through *nous* that we can think of the immediate premise of the true universals.[16]

It is here that Stiegler's insight into the role of mnemotechnics gives a different account of experience. Aristotle's search for an immediate premise of all knowledge still echoes in Husserl's call for the originary givenness (principle of all principles) or Stiegler's idiocy. But Stiegler's analysis of the role of secondary and tertiary retentions in Husserl's description of the 'large now' (*le grand maintenant*) changes the relation of noetic experience to time.[17] Stiegler shows that secondary and tertiary retentions are already at work in Husserl's description of the 'large now' and thus demonstrates the impossibility of the immediate evidence on which Husserl, in the aftermath of Platonic metaphysics, was still building. However, the deconstruction of the possibility of the immediate premise of all knowledge did not do away with the need of the immediate ground of the unity of our experience. After the deconstruction of metaphysics, it has to be found in our need of producing it, that is, in its default. As such, the unity of thought is the unmoved mover that motivates all noetic experience. And since all noetic experience in mortal, living beings supposes the sensitive and nutritive souls, Stiegler will affirm this tendency on these layers too. The difference between these layers is described as an increasing mobility: the nutritive soul or layer only moves as it is assimilating the other within itself, the sensitive soul moves in the community (Leroi-Gourhan's group) and the noetic soul moves within an environment which has become totally mobile, that is, changing. Accordingly, there is no stable unity for the noetic soul at work (*in actualitas*).[18] This conclusion contradicts Aristotle, but is the consequence of the default of *immediate* premises.

Having pointed out that the noetic living being has to produce itself out of a multiplicity of sense-perceptions, Stiegler reconfigures the relation between proper senses and common senses. Aristotle distinguishes between the senses that are proper to the organs of perception, and the senses that are shared by the organs of perception. It is the distinction between, what will be called, the secondary qualities (what Aristotle calls the proper senses) and the primary qualities (what he calls the common senses). Stiegler mentions the embarrassment of the traditional readings, since Aristotle does not mention a specific organ for sensing the common qualities. He points to editorial tricks that suggest a 'common sense', but Aristotle explicitly refuses the possibility of a sixth sense. In fact, there is no need for it, because every sense is also the perception of its specificity, that is, of the community of the senses in which every sense is different from the others. The sense of this community is noetic, says Stiegler, and Aristotle determines it as *logos*. If every sense-perception is simultaneously also the perception of itself, this noetic double is also a reduction. It is here that Stiegler's concept of *idios* gets its sharp edge. Indeed, if sense-perception, as such, is a sense of its community with other sense-organs, or with other sense-perceptions, it is a ratio of what is common between these perceptions. In sensing the community in the differences of these senses, *noēsis* reduces the 'incommensurability of the proper, the idiocy of it, which is also the truth of the intuition, the originary given.'[19] The community of senses gathers the proper sense as 'a tension, an injustice in respect to the idiocy' of it, an idiocy 'which will inscribe itself as a default',[20] that will give way to new idiocies (sense-perceptions) and other thoughts. It is a community of a default of community, Stiegler says.[21]

Thought (*noēton*) on the other hand, the noetic sense, consists in the unity of experience and has also a sense of itself. Thought is always also thinking itself. And thinking itself, it thinks its desire of thinking the unity of all there is. That is why Stiegler says the medium (milieu) of the noetic soul is theo-logical.[22] *Theos* is the unity (*hen kai pan*) of being to which *logos* is tending. It is here that Stiegler's attention to technics performs a twist in Aristotle's theory of the soul. When Aristotle says that the unity of thought is something that has to be produced,[23] he refers to a sense of *technē* that is very different from what Stiegler calls technology. To Aristotle, understanding being is understanding its necessity, whereas *technē* concerns what can be otherwise. Consequently, in a world where everything is perfectly understood, there is 'no room for *technē*, nor in general for human action'.[24] For Stiegler, on the contrary, science has become 'applied technology'.[25] Contemporary science distinguishes itself from both Greek and modern science. Modern science is 'following multiple causal series in nature',[26] finding a place for the contingent action of technics in this multiplicity. In contrast, contemporary science 'utilizes [the causal series], diverts them as a waterway might be redirected by modifying the direction of flow of its course'.[27] And whereas modern technology is only considered to be applied science, contemporary technology opens up a hypothetical-technical horizon which is not that of theorematic light but of hypothetico-technical *making,* sole able to modify processes, and in which one must attempt to orient oneself among the diversity of overabundant possibilities, a diversity which is being aimed at in this *systematic exploration of darkness.*[28]

Defaulting Immanuel Kant's postulates

The shift of techno-science is a shift from theory to praxis, a techno-scientific praxis. The hypothetical-technical horizon of technoscience is the horizon of possibilities that are not guided by the sense of the eternity of being as in Greek thought, nor by the ultimate ground of all beings as in modern thought, but by the horizon of the possibilities which offer themselves to its power. This horizon is not limited anymore by the essential forms of being, or laws of nature, but by the limits of the possibilities of its own performance.

Kant answered the question 'how to orient oneself in thinking?' with recourse to a subjective criterion, that is, the belief of reason. This belief would guide us in the obscure realm of the suprasensible, where our sensibility is unable to provide an object of knowledge. The belief of reason concerns the question of the existence of God, the utmost reality, as ground of the possibility of all reality. To Kant, all understanding is objective; its possibility and its reality are grounded in existence.

But since all insight derives from experience and since we do not have experience of suprasensible things, the use of reason in this realm is only a subjective maxim for orienting ourselves.

Kant's solution is of interest to Stiegler because it allows him to understand that we can explore the realm of the unknown (*inscience*) by sticking to the necessities of the concatenations of reason.[29] Kant conceives this as a need of reason. Stiegler's insight into the technical condition of thought is at once the expression of a not-knowing and the requirement of finding a criterion for orienting ourselves.[30] His double formula, as we saw, is: 'default AS eidos and EIDOS as default'.[31] This difficulty is very different from the one Kant tried to solve. Kant tried to save two things: the unity of phenomena and the possibility of moral action. He accomplished this by crediting reason of being not contradictory with the possibility of the existence of God. Stiegler wants to understand the unity of the flux of the phenomena. He shows that this unity is conditioned by tertiary retentions. But since deconstruction implied an inability to refer to a founding concept of nature, or substance, as in Kant,[32] we have to refer the unity of the flux to the default of our condition, that is, the privation in which we are left after metaphysics.[33] To Kant, this condition is practical, that is, moral. To Stiegler, it is practical because it is technical. The need of reason is the need to produce essences we can share.

Time bars the way to past experience (even our own) and implies that lived experience (*le vécu*) always already composes with the tertiary retentions in which experience has fixated itself. Derrida's demonstration of the fact in *Voice and Phenomenon*, requires that the Kantian solution (the belief of reason) must accompany experience in every step. Indeed, because we cannot access the liveliness of past experiences, we must rely on the possibility of its traces (tertiary retentions) to sustain the unity of consciousness. The unity of consciousness these traces allow is only the promise of an experience (a unity) *to come*. But since these possibilities (Derrida's archi-trace or Stiegler's idiotext), however efficient they are, cannot account for the differences of the experience they render possible (which is always contextual (Derrida) or idiotextual (Stiegler)), this promise is also a requirement.[34] And since our technical condition is implying this *prosthesis*, we have no choice but to respond to it. The belief of reason is the desire to know that is implicit in the promise of the unity of consciousness. In Kant, this promise is the idea of the total determinability of all things. To Stiegler, it is the idea of the total determinability of life.

To Kant the belief of reason requires the idea of an essence of all essences as ultimate ground. It does not require the *existence* of God,[35] only the idea of it as a transcendental ideal. The ideal of this transcendental substratum[36] is the ground of the belief of reason.[37] As such, it is 'the ground [...] to which all thinking of objects in general must, as regards the content of that thinking, be traced back.'[38] In this form, Stiegler could not adopt this belief since essence is missing in our technical condition. Stiegler's reading of the deconstruction of metaphysics seems to leave us with the task of creating an eidetic.[39] Stiegler adopted the transcendental ideal as Kant conceived it. But he could not acknowledge the transcendental substratum of this ideal. The historicity of the task of creating an eidetic would ground in a default that is repeating itself over and over again. Unlike in Kant, the ground of the unity of the phenomena is not the ideal prototype (*Urbild*) of which all consciousness is only a partial approximation. To Stiegler, this ground is to be found in the historicity of one's idiotext, which is never given, but always to come.[40] The idiotext is not the idea of the essence of essences *in individuo* as in Kant, but *in individuo* the essence of what is missing. Kant argued that every negation of an essence (*realitas*) is a privation of this essence, which is presupposing it. This is not the case in Stiegler's idiotext, where default is an absolute negation of essence (default AS eidos), which to Kant cannot be represented (*irrepresentabile*). Stiegler's idiotext is a *nihil negativum* and as such irrepresentable, while the 'feeling of the need which is inherent in reason itself' is a *nihil privativum* of its objective truthfulness (EIDOS as default).[41] In metaphysics, the negation of an essence is the negation of a real possibility (*actualitas in absentia*). In deconstruction, it signifies within the realm of the totality of involvement which is the possibility of impossibility, namely, the mortality of the 'who?' (its temporality). As long as essence has the necessity that characterizes *orthothesis*, it can be shared by other mortals. But that concerns only its possibility, not the possibility of impossibility, which is the horizon of all significance of the 'who?' *in individuo*. And this is why Stiegler had to produce the concept of idiotext. Stiegler produces a concept of idiotext and although it determines the default which it is as an essence (*eidos*), this *eidos* is not the unity of all positive determinations (as in Kant's God). As an essence, it is the unity (*nihil privativum*) of the default of determination (*nihil privativum*) over and over again. This must be thought of *in individuo*, because all negation (all technics) is determining *in concreto*. The unity of idiotext however is an indetermination, the indetermination of life.

We contrast the three postulates of practical reason in Kant (immortality, freedom and God) with the demands of reason (postulates) in Stiegler. In Kant, the postulate of immortality is required because of the necessity of the adequacy of one's own 'duration to the complete fulfilment of the moral law'.[42] That of freedom flows from the necessary presupposition of the 'independence from the world of sense and of the ability to determine one's will according to the law of an intelligible world'.[43] The postulate of the existence of God flows 'from the necessity of the condition for such an intelligible world, in order for it to be the highest good, through the presupposition of the highest self-dependent good, i.e. of the existence of God'.[44] To Stiegler, the highest self-dependent good is idiotext (the who?), which is technical life, freedom is the task of creating an eidetic to share one's idiotext (mortality), and since the unity of the living (phenomena (time)) is dependent on technics, that is, on death, idiotextual life is immortal, as far as technics reach. Stiegler did not publish anything approaching these statements, but these formulations are an attempt to understand his relationship to Kant's postulates in *Technics and Time, 3*.

The soul's composition

Reason is the unchallengeable authority in understanding truth, be it in theoretical, in practical or in technical matters.[45] It is a request for grounding. And although its authority in suprasensible matters has become a subjective criterion, it remains an objective criterion in practical matters. Stiegler adopts Kant's concept of the belief of reason.[46] Kant's grounding of the belief of reason is the result of a critique of metaphysics as a critique of pure reason. It hinges on the gap between form and matter which is bridged in the postulates of empirical thought. Stiegler's answer to the question of the grounding of the belief of reason is a conception of the soul as a noetic sense for the default of our condition.

The power of the intellect is not only a ground and a compass to Stiegler, but also a power that can ground oneself, that is, catch or take hold of someone and leave them powerless, as one says 'being in the hands of the police'.[47] The image speaks very well of the objective *epochē* of our technical condition that can keep us imprisoned in public affairs, while our private understanding is left wanting. It echoes what Stiegler describes in 'The Concept of Idiotext' as public versus privative forms of interpretation.[48] But Stiegler insists that 'it is necessary to *continue* to believe, it is necessary to *want* to believe'.[49] The issue at stake is the very sense of enlightenment: the freedom of speech, that is, the belief in public reason, which was already Kant's endeavour.

Whereas in his doctoral thesis Stiegler based his interpretation of Aristotle's theory of the layering of the soul on Hegel's reading of it, in 'Wanting to believe' Stiegler offers an interpretation that benefits from the developments of *Technics and Time, 3*:

> Aristotle distinguishes three types of souls, that is, of being which find within themselves their movement, their auto-mobility, their animation: the vegetative soul, the sensitive soul and the noetic soul. The sensitive soul inherits the 'vegetativity' of the vegetative soul, and the noetic soul inherits the sensitivity of the sensitive soul (and through this, it equally contains the vegetativity of the vegetative soul).[50]

Traditionally these layers have been associated with the realms of different sorts of living beings: plants, animals and humans. But what interests Stiegler in Aristotle's layering is their composition in humans. Aristotle himself already notes that these levels compose in higher organisms and that the allocation of these forms of liveliness to different parts of the human soul is not self-evident.[51] Aristotle separates the different forms of liveliness (the nutritive, the sensitive, the noetic) in order to understand the way they also work in composition as in desire (*orexis*). And talking about desire he notes there is a special role of the sense of touch without which no animal can survive.[52] It is the sense that enables all animals to furnish themselves with what they need and avoid what threatens them.[53] This is what Stiegler adopts in a technical sense, when he talks about the hand that reaches out or makes a fist, and which is also the instrument of instruments, as Aristotle says, when he is introducing the noetic as such.[54] The tradition has read this image of the hand as a metaphor.[55] Stiegler on the contrary wants to adopt a compositional reading of Aristotle's theory of liveliness, in which each layer has a role. And as Stiegler indicates,[56] *noēsis* is vital to *aisthēsis*, it is clear to him that the sense of touch (the hand) is vital to humans in respect of the nutritive realm.[57] The specificity of the nutritive mode of being is that it relates to the world by assimilating its matter or by rejecting it. According to Aristotle the sense of touch in animals is a sense of belief in its bodily condition.[58] Aristotle thinks that belief (*doxa*) is not limited to inferences, because it also extends to animals that do not possess reason.[59]

Stiegler grounds the belief of reason in the idea of the existence of being in default (i.e. mortality). In fact, he affirms that the principle of reason is made to default,[60] as was already suspected by Aristotle in *Metaphysics Z3*. Kant understood this default as a belief of reason that needs to elevate itself beyond the limits of what it can know, to the idea of the existence of an ultimate ground, the sum of all determinations, God. Stiegler showed however that the technical condition of knowledge means that these eidetic determinations are possibilities that have been produced (creations) as the unity of the phenomena necessary to orient ourselves. In order to share this unity with others we need to master its technical condition. We are able to do this only intermittently, as the actualization of possibilities we are thrown in. These technical possibilities do not, as Kant thought, find their limits in the possibility of noetic determinateness (knowledge), but in the possibility of the impossibility of the 'who?', that is, in the limit of its liveliness as such (which is an indetermination). This is why Stiegler is interested in adopting Aristotle's theory of liveliness. In his essay, Stiegler reads this layering in the light of Aristotle's notion of potentiality. And however heterodox his interpretation of *De Anima* may be,[61] it is clear that he adopts Aristotle's view of the cumulative structure of this layering. He characterizes this cumulation as an inheritance that is determined through the *semeiōsis* of the noetic, which includes discernment (*krinein, diagnōsis*) and judgement (making-a-difference). This interpretation of the layering of liveliness could be understood as the traditional (metaphysical) understanding of the *zōon politikon*, that is, as '*un supplément d'âme*', a *logos* that would 'give spirit to the animal's soul'.[62] But of course Stiegler rejects this. When he uses the word 'inheritance' he refers to his elaboration on tertiary retentions. 'Inheritance' designates a dynamic that is very different from the metaphysical understanding of liveliness, where life is understood as a form which in its perfection is pregiven (Plato) or actual (Aristotle). Since the perfection of these forms is conceived as an identity that pervades in the diversity of all its occurrences, the living being that possesses the faculty of being moved by the very perfection of these forms is most akin to eternal life. Stiegler would not deny that if they want to create an eidetic, humans should be moved by perfection. But since this perfection is not given, it must be produced and its products (idealizations) can only be shared as tertiary retentions – which is 'death [seizing] hold of life'.[63] The creation of an eidetic articulates a dynamic of life and death. All this however could still lead to the same metaphysical construction. The dialectic of life and death is already present in the first metaphysical text, in *Phaedo*. However, the articulation of life and death in our technical condition is not a dialectic. It is a retention, an objective *epochē* and with it the danger of proletarianization and a promise. Deconstruction is indeed a repetition of Socrates's exercise in dying.[64] But whereas Socrates conceives life and death to be an opposition that dialectics can surmount in favour of an eternal life, deconstruction shows death in life (the possibility of the impossibility) is the horizon of human existence.

The wound

Death may seize life, but for Stiegler it is the living hands that seize death. It is the living that seizes the promises of death, to repeat, carry them further and renew them, that is, also to repeat them. It is clear that this philosophy, inspired by Nietzsche, is an affirmation of life. And this affirmation is supported (*porté*) by the instrument of instruments, the hand. It is the hand that can seize the occasion, that is, the chance of the singularity which life offers. Stiegler says singularity 'is *clearly* this chance'.[65] However, if singularity, this '*unconditional* desire, which it is *impossible to renounce*',[66] is to guide deconstruction, we must ask: when did singularity become a common goal, and to what extent is it common in other cultures, other cosmotechnics? And what does it mean to qualify singularity as the object of a desire, as a desire to desire?[67] Socrates talked about the singularity of his place,[68] which he could not desert because he did not own his life proper, but was responsible for it before God. For Socrates, his own singularity was dependent on God, being determined as the possibility of an *ousia* (presence) which should remain unaffected when separated from contingency. The deconstruction of metaphysics however showed that an answer to the question of our singularity would require an understanding of how the horizon of temporality already determines the possibilities of answering this question. Existential analytics of being-there showed this horizon to be mortality. Stiegler understands this possibility as the technicity of our existence. The singularity of what one shares and what is missing in the tertiary retentions one has at hand, is the objective *epochē* of the pre-individual milieu in which we live. This milieu is caught in the horizon of death. The singularity of the moment requires to reinvent the eidetics of which the tertiary retentions are but traces in order to account for the altered possibilities of the impossibility of being-there. And since tertiary retentions do not solely concern noetic forms, but all ways of living and doing,[69] the objective *epochē* is a default that not only deprives the spirit, but also the flesh: 'Singularity, which is also called *idios,* is first of all a wound. It is a wound of the flesh that forms a defect [*qui se fait défaut*]. But one that is necessary.'[70]

The wound is the event which gives birth to the singularity of what happens to us. In referencing Joë Bousquet's notion of the wound, Stiegler wants to adopt something of the counter-actualization, conceived by Deleuze in *The Logic of Sense*. In Stiegler's reading, Deleuze's actor has become a man of action, even '*un homme de main*', a henchman. When he is appropriating the default of his own condition (his mortality), he must double it, in a desperate effort to embody it. This is the sense in which Deleuze can say with Bousquet '[m]y wound existed before me, I was born to embody it.'[71] The temporality of the event is that of the instant. But the event is imprisoned, like Stiegler's 'who?' in the objective *epochē* of tertiary retentions. 'To the extent that the pure event is each time imprisoned forever in its actualization, counter-actualization liberates it, always for other times.'[72] The instant is the point where the will articulates a sense of the wound that is incomparable although it is destined to be repeated. Where Deleuze talks about 'a sort of leaping in place *(saut sur place)* of the whole body which exchanges its organic will for a spiritual will',[73] Stiegler would say: 'it is a matter of *thinking the relation between the will and the hand, such that they are believing. That is, also, intellectual.*'[74]

We indicated already that Stiegler characterizes the interaction of the different layers as a symbolization (*semeiōsis*) and an inheritance, in which higher layers inherit the potentiality of the lower, in a kind of elevation that eschews traditional metaphysics. Inversely, the higher layers pervade all of the lower ones, as a discernment fixated in an expression.[75] This discernment, which is noetic, can occur on all levels. For the living being having discernment, the nutritive or the sensitive concerns as much the noetic sense as does a mathematical equation. Stiegler proposes to articulate these interactions between the layers of the soul using the distinction between potentiality and actuality.[76] This seems to make no sense in relation to the major part of *De Anima* where each level is characterized as a potential with an actualization of its own. However, when Stiegler is talking about the interaction of the layers he talks from the practical point of view (*praxis*) that is discussed at the end of *De Anima*, where Aristotle talks about *orexis*, desire, as that which produces local movement as productive of locomotion.[77]

When Aristotle queries the principle of animal locomotion, he touches on the question of desire. Having discarded the nutritive, sensitive and noetic layers, he answers that it must be desire that moves living beings locally,[78] since all other local movements seem to be the result of forces alien to these living beings.[79] If an animal is capable of desire, then it can move itself.[80] Further, the object of desire is 'either the good or the apparent good'.[81] Desire requires imagination and imagination 'is either concerned with reasoning or perception'.[82] Desire however does not require belief, because belief implies deliberation, hence reason, and desire can move without deliberation, for example, in a wish. Although, Aristotle writes that 'by nature the higher is always predominant and effective',[83] when concluding his analysis of the soul, the question of what is necessary for a living being casts a different light on the analysis. Suddenly, the nutritive layer of the soul comes to the fore, as the orientating force of all living beings, since everything 'that lives and has a soul must have the nutritive soul'.[84] This applies as much to all living beings in as much as they have bodies. For example, if sensitive souls are to survive, they primarily need touch, as 'the animal is an ensouled body, and every body is tangible'.[85] Touch is key to the orientation of all animals, because it is the only sense that shares the bodily condition of the animal ('For the other senses, smell, sight, and hearing, perceive through other things.')[86] In 'Wanting to believe' Stiegler claims that 'insofar as action belongs to *aesthesis, nous* is itself aesthetic'.[87] But we should draw the line even further, down to the nutritive level, as Aristotle does. We should try to understand that if the flesh or the body is the medium of sensibility,[88] and *noēsis* a form of sensibility, then the body (*soma*) must be the medium of all noetic animals. This is why Aristotle can say that the hand is the organ of organs. And this is why Stiegler can identify the hand with the flesh.[89]

The organ of organs

To elucidate this, we propose to understand this figure through a thought experiment. If we would try to think of the soul as being a hand it is obvious that coming to grips with matter would be its vegetative or nutritive sense, its sensitive mode would be touching and its noetic sense conception. To Stiegler, the hand (the flesh) is not only an instrument, but the name of the living insofar as it is exposed to its technicality 'insofar as it is a deadly affair, and insofar as it makes death [*fait le mort*]'.[90] That is the logic of all technics (tertiary retentions). Even when it makes one live, it is at the cost of the death of all it takes to do it (entropy). But if the hand is the expression of the liveliness of the living being what can it mean to 'have a hand in dying'? Can one want to have a hand in one's own death? Socrates thought he should not have a hand in his own death,[91] although he was convinced that philosophy was an *ars moriendi*.[92]

To Aristotle the principle of the hand is its substance, it is a composite of form and matter, independent and being what it was over time (*to ti en eīnai*).[93] In its deconstructed form this principle is not a given anymore but the formula of a repetition, that is, the return of a question within the horizon of our existence. What in metaphysics is conceived as the actuality of an identity becomes, in its deconstructed form the possibility of an impossibility: its reproduction. Can you *want* to have a hand in your own death if the hand is the name of the living? In a sense we do nothing but this. If, following Stiegler's teaching, the living is dependent on its technicality, it seems that having a hand in one's own death would be the very condition of the unity of the living and its technical immortality. If we need to create the eidetic that we share with others, by repeating the one we inherit, that is, by differentiating it, freedom would be the possibility of sharing a belief in the promise that we determine or cause our lives to be what they are. But if the hand exposes us to our own technicity, it seems that every act of freedom indebts the future and reifies it, confiscating it within a projection.[94] On this account, it would seem that we would have to answer the question of the *Phaedo* anew: is there life in death? In our technical condition the answer must, more than ever, be sought on this side of the Acheron. The hand of death may seize life and vice versa. But in this deconstructed scene, the solution will not be found in the *idea* of life, which was the dialectical solution in *Phaedo*.[95] To Stiegler, the *metabolē* of death into life and life into death is a pharmacology, an opposition of forces *in materia*. And although both solutions claim a life that surmounts death by embracing it, they fundamentally differ; 'Wanting to Believe' offers a new approach to this old question. The deconstructed form of immortality is a form of idiosyncrasy of the flesh in a combat against entropy – a differing.

What happens when the hand takes its own life, according to Stiegler? What happens in the soul of the henchman? It is not only the power of its grip, the sensitivity of its touch or the conception of what it has at hand that goes astray. It is none of these layers alone. Neither is it the capacity to handle these things in a grammatized form. As organ of organs, it is relinquishing the belief of the living body (soul) that aims at sharing its idiosyncrasy. There is a privation in Stiegler's pharmacology that is not logical, but material. The relinquishing of the hand may be gradual, but it ends with an absolute negation that is not possible in dialectics, be it idealistic or materialistic.[96] The soul can linger a while, but in the end, it cannot be sustained. It is the possibility of a hand that can lose its grip, burn its fingers or fail to put things together anymore.

Notes

1. We will use the italicized *Technics and Time* to refer to the overall writing project of technics and/as time.
2. Bernard Stiegler, La technique et le temps 1: La faute d'Épiméthée – 2: La désorientation – 3: Le temps du cinéma et la question du mal-être suivis de Le nouveau conflit des facultés et des fonctions dans l'Anthropocène, Paris, Fayard, 2018, 10.
3. Information from Jean-Hugues Barthélémy.
4. Bernard Stiegler, 'Ce qui fait défaut', *Césure* 8 (1995): 231–78.
5. The text published in *Intellectica* is older (1985) than the one published in *Césure* (1995). See information from Jean-Hugues Barthélémy in Pierre Steiner Pierre & John Robert Stewart, 'Présentation des textes', *Intellectica*, 53–54/1–2 (2010): 41, DOI available at: https://doi.org/10.3406/intel.2010.1177.
6. Apart from these one would have to consider the notion of stupidity (*bêtise*) which is closely related to idiocy. We will not do it here.
7. Bernard Stiegler, *La faute d'Épiméthée: La technique et le temps*, Paris: École des Hautes Études en Sciences Sociales, 1993, doctoral thesis, 559.
8. Stiegler, *La faute d'Épiméthée*, 559.
9. Jacques Derrida, *Khôra*, Paris: Galilée, 1993.
10. Aristotle, *De Anima, Books II & III*, trans. David W. Hamlyn, Oxford: Clarendon Press, 1993², 20–1 (416b), 'digesting'. For references without translation we use the Bekker pagination.
11. Stiegler, *La faute d'Épiméthée*, 560.
12. Stiegler, *La faute d'Épiméthée*, 425–6; same text in Bernard Stiegler, *Technics and Time, 2: Disorientation*, trans. Stephen Barker, Stanford: Stanford University Press, 2009, 156. We prefer translating '*comme*' with 'as' instead of 'qua'. Also, we kept the same formatting as the text in *La faute d'Épiméthée*.
13. Aristotle, *Posterior Analytics*, trans. Geoffrey R.G. Mure, in William D. Ross (ed.), *The Works of Aristotle*, Oxford: Clarendon Press, 1925, William D. Ross, 323 (99b).
14. Aristotle, *Posterior Analytics*, 100b, '*epagōgēi*'.

15 Aristotle, *Posterior Analytics*, 99b-100a.
16 Aristotle, *Posterior Analytics*, 100b.
17 Bernard Stiegler, *Technics and Time, 1: The Fault of Epimetheus*, trans. Richard Beardsworth & George Collins, Stanford: Stanford University Press, 1998, 246-8.
18 There is, according to Stiegler, a metastability of the noetic soul, but this concerns the dispositions of the noetic soul (*hexis*), not her actualization.
19 Stiegler, *La faute d'Épiméthée*, 565.
20 Stiegler, *La faute d'Épiméthée*, 565.
21 Stiegler, *La faute d'Épiméthée*, 563; see also Stiegler, *Technics and Time, 2*, 243.
22 Stiegler, *La faute d'Épiméthée*, 559. He will distinguish his conception of *theos* from Aristotle's conception, following Kant's belief of reason. See Stiegler, *La technique et le temps 1-2-3*, 804.
23 Aristotle, *De Anima*, 430b.
24 Pierre Aubenque, *La prudence chez Aristote*, Paris: Presse Universitaire de France, 1997^2, 68.
25 Bernard Stiegler, *Technics and Time, 3: Cinematic Time and the Question of Malaise*, trans. Stephen Barker, Stanford: Stanford University Press, 2011, 189.
26 Stiegler, *Technics and Time, 3*, 69.
27 Stiegler, *Technics and Time, 3*, 190.
28 Stiegler, *Technics and Time, 3*, 191, translation modified.
29 Immanuel Kant, 'What is orientation in thinking?', in Immanuel Kant, *Political Writings*, trans. Hugh B. Nisbet, ed. Hans S. Reiss, Cambridge: Cambridge University Press, 1991, 235-49, 240.
30 Stiegler, *Technics and Time, 3*, 169.
31 'Yet it could be that this default is here, *as such*, still (a) reason, simultaneously a motif and a necessity-in fact, a motif and the necessity of the incalculable, including *death* itself, as what un-determines *Dasein* and is the human (living) being's *great default* (but also precisely the element of chance-in-life, the principle of an immense process of individuation we call "evolution")', Stiegler, *Technics and Time, 3*, 192. What Stiegler expresses here in a conditional mode, is in fact the horizon of his proper thought.
32 Stiegler, *Technics and Time, 3*, 199, 203.
33 But to Stiegler this is already the privation of the Zinjanthrope using a biface.
34 We have to take into account the promise the tertiary retention contains (i.e., the interpretation of which is a trace). Taking into account this promise is first of all acknowledging that there was a promise at work here.
35 Immanuel Kant, *Critique of Pure Reason*, A580. We use the translation of Paul Guyer and Allen W. Wood, Cambridge: Cambridge University Press, 1998. For reference we give the reference page of the German edition of 1781.
36 Kant, *Critique of Pure Reason*, A575; which is all of reality (*omnitudo realitatis*).
37 '[W]hat I call the ideal, by which I understand the idea not merely *in concreto* but *in individuo*, i.e., as an individual thing which is determinable, or even determined, through the idea alone', Kant, *Critique of Pure Reason*, A568. It is the representation of a muster (*Urbild*; *prototypon transcendentale*) that cannot be specified, 'because apart from experience one is acquainted with no determinate species of reality that would be contained under that genus' (Kant, *Critique of Pure Reason*, A577). It is an ideal for orienting ourselves that allows us 'to assess and measure the degree and the defects of what is incomplete' (Kant, *Critique of Pure Reason*, A569-570).
38 Kant, *Critique of Pure Reason*, A576.

39 For Stiegler, this task is a technical-practical task. Its paradigm is that of technoscience, which is the contemporary figure of *epistēmē*. It confronts us with the impossibility of having recourse to a concept of the nature of things (e.g., biotechnology, Stiegler, *Technics and Time 2*, 155 ff.). Such a concept would require an idealization in the Husserlian sense. But Stiegler's analysis of primary, secondary and tertiary retentions prevents us from referring the unity of the flux of phenomena to the ideal of a living present (*lebendige Gegenwart*). The consistence of the phenomena can only be projected through the idealizations enabled by the *orthothesis* of the tertiary retentions in which this unity is being fixated. But since the unity of the flux of phenomena cannot refer to an invariant nature, or substance, but only to the fixation of its sedimentation, the request for the fixation of the unity of the flux of phenomena in another tertiary retention is bound to be repeated again and again. The horizon of this request is the default of reason itself (Stiegler, *Technics and Time 3*, 192, 201). On the question of the creation of an eidetic, see Jacques Derrida, *Edmund Husserl's Origin of geometry: An Introduction,* trans., with a Preface and Afterword by John P. Leavey Jr., Lincoln and London: University of Nebraska Press, 1989.
40 The 'who?' as he designates it f.i. in Stiegler, *Technics and Time, 3*, 174.
41 Immanuel Kant, 'Attempt to introduce the concept of negative magnitudes into philosophy (1763)', in Immanuel Kant, *Theoretical Philosophy 1755–1770*, trans. and ed. David Walford & Ralf Meerbote, Cambridge: Cambridge University Press, 1992, 203–42, 211; Kant, *Political Writings*, 240, 244.
42 Kant, *Critique of Pure Reason*, 168.
43 Kant, *Critique of Pure Reason*, 168.
44 Kant, *Critique of Pure Reason*, 168.
45 Kant, *Political Writings*, 243: 'the ultimate touchstone of the reliability of a judgement'; see also Kant, *Political Writings*, 249.
46 Stiegler, *Technics and Time, 3*, 200 ff., 180 ff.
47 Bernard Stiegler, *The Decadence of Industrial Democracies. Disbelief and Discredit, Volume 1*, trans. Daniel Ross & Suzanne Arnold, Cambridge: Polity Press, 2011, 132.
48 Bernard Stiegler, 'Le concept d'"Idiotexte": esquisses', *Intellectica: Revue de l'Association pour la Recherche Cognitive*, 53–54/1–2 (2010): 51–65, 57, DOI available at: https://doi.org/10.3406/intel.2010.1178.
49 Stiegler, *The Decadence of Industrial Democracies*, 132.
50 Stiegler, *The Decadence of Industrial Democracies*, 133; 'Wanting to Believe: In the Hands of the Intellect' is the 4th and final chapter of this book.
51 Aristotle, *De Anima*, 432a22–432b27 and 433a31–433b5.
52 Aristotle, *De Anima*, 435b.
53 Aristotle, *De Anima*, 434b: 'Since the animal is an ensouled body, and every body is tangible, and it is that which is perceptible by touch which is tangible, the body of an animal must also be capable of touch, if the animal is to survive. For the other senses, smell, sight, and hearing, perceive through other things, but anything which touches things will be unable, if it does not have sense-perception, to avoid some of them and take others. If that is so, it will be impossible for the animal to survive'.
54 Aristotle, *De Anima*, 432a.
55 Stiegler, *The Decadence of Industrial Democracies*, 154.
56 Stiegler, *The Decadence of Industrial Democracies*, 136.
57 Stiegler, *The Decadence of Industrial Democracies*, 159, on consumption and Georges Bataille's consummation.
58 Aristotle, *De Anima*, 435a15–19 and 434a7 'something bodily'.

59 Aristotle, *De Anima*, 434b on imperfect animals.
60 Stiegler, *Technics and Time, 3*, 182.
61 As he says himself; Stiegler, *The Decadence of Industrial Democracies*, 137.
62 Stiegler, *The Decadence of Industrial Democracies*, 136.
63 Stiegler, *The Decadence of Industrial Democracies*, 147.
64 Plato, *Phaedo*, 64a.
65 Stiegler, *The Decadence of Industrial Democracies*, 159.
66 Stiegler, *The Decadence of Industrial Democracies*, 138.
67 Stiegler, *The Decadence of Industrial Democracies*, 162.
68 Plato, *Phaedo*, 62a-c.
69 Stiegler, *The Decadence of Industrial Democracies*, 133, 176-7 fn. 53.
70 Stiegler, *The Decadence of Industrial Democracies*, 160.
71 G. Deleuze, *The Logic of Sense*, trans. Mark Lester with Charles Stivale, ed. Constantin V. Boundas, New York: Columbia University Press, 1990, 148.
72 Deleuze, *The Logic of Sense*, 161.
73 Deleuze, *The Logic of Sense*, 149.
74 Stiegler, *The Decadence of Industrial Democracies*, 132, modified.
75 Stiegler, *The Decadence of Industrial Democracies*, 133.
76 See. Stiegler, *La faute d'Épiméthée*, 522.
77 Aristotle, *De Anima*, 433a; immediately before this he briefly discusses the difficulty of determining the interaction of the layers he distinguished, 432a22-432b6 and 433a31-b5.
78 Aristotle, *De Anima*, 433a.
79 Aristotle, *De Anima*, 432b.
80 Aristotle, *De Anima*, 433b.
81 Aristotle, *De Anima*, 433a (trans. Hamlyn, 70).
82 Aristotle, *De Anima*, 433b.
83 Aristotle, *De Anima*, 434a.
84 Aristotle, *De Anima*, 434a.
85 Aristotle, *De Anima*, 434b.
86 Aristotle, *De Anima*, 434b.
87 Stiegler, *The Decadence of Industrial Democracies*, 136.
88 Stiegler seems to apprehend this, but the whole text is a demonstration of the primordiality of the hand, that he explicitly identifies with the flesh: 'first of all because it has hands. That is, flesh', Stiegler, *The Decadence of Industrial Democracies*, 155.
89 Obviously, we should not forget that the main contribution of Stiegler to our conception of the body proper is its composition with exosomatic bodies (in line with the teaching of Georges Canguilhem). But that should not make us forget that in a sense the exosomatic, as it is conceived as part of the body proper, is in turn becoming endosomatic. But to Stiegler, the distinction between inside and outside does not coincide with the milieu of touch, as in Aristotle (*De Anima*, 434b), but between the accidentality and automobility of the *idios*. Concerning the nutritive layer, we may have to consider the following: Stiegler wants to think the layering of the soul as a composition. That implies that the different levels influence one another. Stiegler always envisions the effects of the exosomatic into the noetic. In 'Wanting to Believe' he considers these effects as an inheritance (Stiegler, *The Decadence of Industrial Democracies*, 133) and from the point of view of action (Stiegler, *The Decadence of Industrial Democracies*, 136). Nevertheless, if his does speak of a becoming aesthetic

of the noetic, he does not speak about a becoming nutritive or vegetative of the noetic, although he clearly includes not only the sensitivity of the hand but also its flesh in this becoming active of the noetic. In fact, this extension of the nutritive to the exosomatic milieu is already present in Aristotle's conception of the nutritive soul. The nutritive soul is an automobility that assimilates the exosomatic into the endosomatic. And, as we remarked in fn. 88, in fact the whole text of 'Wanting to Believe' seems to be a plea in favour of such a conception. Nevertheless, this extension would imply specific problems that have to be addressed, specifically the problem of how the self that is noetic and nutritive in one, is not intrinsically *autological*.

90 Stiegler, *The Decadence of Industrial Democracies*, 155.
91 Plato, *Phaedo*, 62c.
92 In contrast, Aristotle, as a physician, thought more about having a hand in one's health and it is his preferred example when defining technics.
93 In *De Anima*, Aristotle speaks of the essence (*ousia*) of the eye which is its soul, that is, sight (*opsis*, in Aristotle; *De Anima*, 412b20–21). Stiegler uses the plural in the expression 'the hands of the intellect'. His deconstruction of metaphysics implies that it is technics, here technics of the hands plural, which is the principle of the intellect singular. We try to address this question of unity and plurality in our *Postscript to Stiegler's Hands* (forthcoming).
94 It is Heidegger's '"Being-the-basis for a Being which has been defined by a 'not'" - that is to say, as "Being-the-basis of a nullity"' (Martin Heidegger, *Being and Time*, trans. John Macquarrie & Edward Robinson, New York: Harper and Row, 1962, 329) The 'not' in Stiegler's worldhistorical reinterpretation of it is the determination of the tertiary retention as the retention of what is not possible anymore, i.e. the lively experience whereof it is but the trace.
95 Rudolf Boehm, *Ideologie en ervaring*, Kritiek, Gent, 1984, 32–9, available at: https://a03834fb-d85e-47df-a755-570a10a67c74.filesusr.com/ugd/58f4da_cd88881d461347f085da96d8dd8d93c2.pdf.
96 See Paul Willemarck, 'De indrukken: Over het onaanraakbare in Boehms *Schijnbare werkelijkheid*', in *Wat moet? En wat is nodig?*, ed. Paul Willemarck, Antwerpen & Apeldoorn: Garant, 2018, 173–215, 187–8.

Part III

THINKING *DIFFÉRANCE*: LIFE, TECHNICS, EPOCHALITY

Chapter 8

NEGENTROPY AND *DIFFÉRANCE*: STIEGLER'S
MEMORIES OF THE FUTURE

Georgios Tsagdis

Late in his life, Bernard Stiegler begins to articulate the totality of his thought and his project around the ramifications of entropy. Entropy and its correlate negentropy come to the fore in such a manner that they may be seen to provide the coordinates of the third phase of Stiegler's work, the period that Daniel Ross calls *neganthropological*, following the *technological* and the *organological-pharmacological* phases.[1]

The concept of entropy weaves together the fundamental laws of thermodynamics and in a schematic, preliminary manner can be understood as a designation of the amount of maximally distributed or 'disordered' energy in a system, the energy of the system that is thus irreversibly unavailable for the production of any kind of work, or indeed any kind of play. Its correlate opposite, negative entropy or *negentropy*, designates accordingly – and even more schematically – a systemic energy differential that subsists in states 'ordered' in such a manner that work can still be effected.

It is in entropy, and in specific, its role in the second law of thermodynamics, where time, as a physical variable, emerges as a constitutive element of reality. Among all fundamental equations of physics, only the second law postulates irreversibility. Thus, time is often identified, or very nearly so, with entropy, an elision questionable in itself and doubly questionable as soon as consciousness is superimposed on the two, making in effect entropy the explanation of the inability of consciousness to remember the future. If however, 'it's a poor sort of memory that only works backwards,' as the White Queen tells Alice in Wonderland, and if one wishes to craft a memory for the future, to craft that 'genuine repetition' which for Kierkegaard consists in being 'recollected forward,'[2] it is perhaps most conducive to resume labour at the point of the untimely interruption of Stiegler's work and its promise, a point marked by the eruption of entropy onto the foreground of his thought.

Perhaps the best way to begin this labour is by questioning what has premised these opening reflections, questioning, that is, whether one is warranted in speaking of an 'eruption' of entropy, in the strong sense of a new conceptual formation.

Surely, if Stiegler is taken first and principally as a thinker of Promethean fire, as soon as one asks on the reasons that led him to foreground entropy in recent years, one is called to trace the place and function of the concept in his earlier work and assess whether this late emphasis can be understood as marking a shift significant enough to demarcate a distinct conceptual space. Indeed, if late Stiegler's hyperbolic charge that 'the denial of the pre-eminence of the thermodynamic question after Clausius is common to most of those who have tried to overcome metaphysics, that is, passive nihilism', among whom Stiegler includes Marx, Nietzsche, Heidegger and Derrida,[3] and if this denial is identified with the denial of the question of the *pharmakon*, which for Stiegler weaves the history of metaphysics,[4] it seems imperative that prefigurations be sought, in order to map the overall trajectory of the thermodynamic question across Stiegler's work. Although this essential undertaking cannot be carried out here, a cursory survey of the *Technics and Time* trilogy should suffice to set the remainder of the analysis on firm ground.

Technics and Time, 1, appearing in 1994, makes reference to thermodynamics, entropy and negentropy in four distinct passages. In the first, the industrial revolution is cast as a thermodynamic revolution, while innovation is taken to rest on and proceed from investment out of available capital,[5] a clear parallel with Gibbs free energy, a requisite for all negentropic configurations, a parallel which Stiegler however does not explicate.

In the second passage, Stiegler parallels technical progress to 'negentropy', the first mention of the term in his work, which he proceeds to designate as an 'ineluctable complexification of genetic combinations,' an increase in complexity which allows no regression.[6] The third passage, following closely upon the second, marks 'technical dynamism' as 'irresistible negentropy'.[7]

These three passages outline from the outset the parameters of Stiegler's understanding of the thermodynamic polarity. The industrial revolution does not merely occasion the discovery of a physical substratum of reality, but fashions and thrusts this reality onto the historic stage. Here, the preciousness of a 'for the first time' gains traction, as the conditions of history and the unfolding of its narrative begin to converge. As agent of this development, technics is already clearly understood as both entropic and negentropic, in tandem with the ambivalence of the biological.

This becomes particularly palpable in the fourth passage, in which Stiegler relates the optimism of the eighteenth century, its faith in the liberating potential of the emerging thermodynamic machines, soon frustrated by the incorporation of the human as their appendage. The twentieth century renews the liberating promise of the thermodynamic revolution with the creation of cybernetic machines 'capable of producing negentropy'.[8] Technics, it is already clear, represents an entropic fear and a negentropic hope. Stiegler proceeds:

> More profoundly than the relinquishment of the human's place as technical individual beside the machine, the threat of entropy makes possible the anguish in which the human experiences technical evolution. Against this, optimism is justified through reference to a thought of life, because technical evolution appears as a process of differentiation, creation of order, struggle against death.[9]

Life is here already the counterpoint, the locus of hope from which the negentropic potential will have to be drawn. But the first volume of *Technics and Times* will go no further.

Technics and Time, 2, two years later, will make a passing reference to the diagnosis of the first volume, the diagnosis, that is, of the thermodynamic revolution effecting the mobilization of decontexutalized capital, along with the concomitant and necessary development of an infrastructural network of information.[10] More importantly, Stiegler warns here that unless the *who* of humanity is addressed with a new answer within the *what* of technics, 'entropy seems unavoidable as the world waits for catastrophic white noise coming to parasitize the system through its own vulnerability'.[11] In order to make out of the anticipation of the future something other than the catalyst of a white noise apocalypse, Stiegler calls for a political economy of memory.

By 2001 and the publication of *Technics and Time, 3*, the references to entropy and negentropy proliferate, appearing often not as corollaries of the thermodynamic revolution and its legacy, but as autonomous operative concepts. For instance, while Stiegler is prepared to recognize the creation of a 'literary public' in the Enlightenment as a moment of negentropy, he believes that the promise of this moment has been overtaken and obliterated by hyper-digitalization and the industrialized standardization of mass consciousness resulting in 'a homogeneous entropic soup', a 'septic tank', in which the spirit decomposes.[12] This entropic standardization of consciousness is further aggravated by 'the entropy of consumption' which for Stiegler aims at 'self-cancellation, at nullity, at nothing'.[13]

Unsurprisingly, the entropic effects of late capitalism engulf culture as well by disturbing the very coordinates of life what Stiegler calls cardinality and calendarity. Stiegler observes that 'all human beings live the experience of their cultural singularity as a gauge of "vitality" (of negentropy)',[14] and decries the disappearance of that each time unique experience, the disappearance of singularities. The potential for individuation or 'subjectivization' is most violently undercut by calculability which results in 'generalized entropy',[15] nothing less than the eclipse of the future, the eclipse of an '*adoptable to-come*'.[16]

In order to counter this bleak prospect, Stiegler calls for the fashioning of 'negentropic difference', that is, of relationships that exceed oppositionality and rest on the distinction of their terms which are themselves produced through the relationship. This negentropic difference, which is of paramount methodological significance in this reconstruction, Stiegler qualifies as 'necessarily dynamic, activating the composing – without confusion – of the *who?* and the *what?* of the probable and the improbable, the synchronic and diachronic, calculation and [the] undetermined, perception and imagination, *I* and *We*, past and future, future and to-come.'[17]

There is little in the above summary that is not taken up and carried forward by Stiegler in recent texts, such as the *Age of Disruption* (2016), the collection *Neganthropocene* (2018), the *Nanjing Lectures* (2020) or the collective text *Bifurquer* (2020). Although these passages are scattered across the hundreds of pages of *Technics and Time*, not only is there no renunciation and no break of thought, but nearly all the key elements that fashion Stiegler's late understanding of entropy and negentropy seem already in place. Not only is the thermodynamic question not an afterthought for Stiegler, but unthematized and largely unsupported as the character, relation and function of entropy and negentropy remain in these early writings, it is clear that their evocation is anything but incidental. (Neg)entropy is from the outset the necessary supplement.

Something does change however, something more than the density and frequency of the references to entropy and negentropy. Although an event is always plural, arriving both before and after itself, one may localize a decisive moment of this evolution in *Automatic Society, Vol. 1*, published in 2015.[18] In its conclusion, Stiegler confronts what he considers of primary significance in the conclusion of *Tristes Tropiques*. He quotes Claude Lévi-Strauss:

> From the time when he first began to breathe and eat, up to the invention of atomic and thermonuclear devices, by way of the discovery of fire – and except when he has been engaged in self-reproduction – man has *done nothing other* than blithely break down billions of structures and reduce them to a state in which they are no longer capable of integration.[19]

Although Stiegler breaks off the quotation, the passage continues:

> taken as a whole, therefore, civilization can be described as a prodigiously complicated mechanism: tempting as it would be to regard it as our universe's best hope of survival, its true function is to produce what physicists call entropy: inertia, that is to say. Every scrap of conversation, every line set up in type, establishes a communication between two interlocutors, levelling what had previously existed on two different planes and had had for that reason a greater degree of organization. 'Entropology', not anthropology, should be the word for the discipline that devotes itself to the study of this process of disintegration in its most highly evolved forms.[20]

It is this, '"entropology" not anthropology,' that seems to weigh most heavily on Stiegler and spurs him to proclaim a new philosophical anthropology, a 'neganthropology'. The term emerges thus not as an idiosyncratic neologism, but as a direct response to what Stiegler considers as Lévi-Strauss's pessimism. Stiegler disregards the – admittedly troubled – 'And yet I exist,' with which Lévi-Strauss moves on from the above passage,[21] since for Stiegler 'entropology' designates the announcement of an imminent arrival of the 'end times' and the annihilation of the time that remains until this arrival, the reduction of this time to a mere nothing, a becoming without future.[22] The principal mark then of this third phase of Stiegler's work, is its articulation as a generalized project of the 'interim' or the 'inter-esse', a generalized project of the time that remains. Provocative as the intimation may appear at first, the principal concern, the interest and care in which this phase sums itself up, is the fashioning of a negentropic techno-bio-thermodynamic *katēchon*, or rather of a plurality of such local *katēchonta*.

Although this claim cannot be here pursued further, the *neganthropological* project's affinity to the politico-theological logic is not limited to resonant overtones. Stiegler explicitly attempts to dissociate the nomos of the earth from its Nazi legacy in order to think the eschatological moment of algorithmic capitalism. He declares, in the first person: 'I posit that Schmitt, by assigning *nomos* to the locality of a plot of land and its borders, unwittingly raises the inherent problem of *locality* involved in any negentropic bifurcation, in life'.[23] Moreover, even though it must be remembered that capitalist eschatology is that of a radical finitude, that on the obverse side of the time that remains, there is only white noise, the force of the *katēchon* is operative from the outset – if not earlier. It is already there, in Freud's *Beyond the Pleasure Principle* (1920) and it is unmissable in the first articulation of negentropy in Schrödinger's *What Is Life?* (1944). As Schrödinger observes that 'the essential thing in metabolism is that the organism succeeds in freeing itself from all the entropy it cannot help producing while alive',[24] that the organism expends thus a constant effort to stay 'aloof' of a fate which it is itself constantly precipitating and that it does so, not according to coarse human, that is *dynamic* mechanics, but according to 'the Lord's *quantum* mechanics',[25] the intersection of biology and theology becomes all too apparent. The territory of the body, internal and external to the skin, is the locus of this intersection.

In its remainder, the essay outlines two interrelated aspects that structure the *neganthropological* project and which present themselves as open questions, as memories for the future: the question of *différance* and what one may call the question of 'translation'.

From his first and up until the last of his writings, Stiegler thinks through *différance*; however, within the *neganthropological* project the term assumes a new configuration. Importantly, *différance* does not designate the difference between entropy and negentropy, but the improbable possibility of negentropy within a perpetual entropic becoming. With regard to the time that remains, *différance* designates an unpredictable future (*avenir*), what Heraclitus calls *anelpiston*: the unexpected and unhoped-for occurrence within an entropic becoming (*devenir*), in which 'everything will [inescapably] return to dust'.[26] This is nowhere thematized as a 'Boltzman fluctuation' or 'recurrence', that is, as the rare and outstanding return of the same singularity, the same arrangement of microstates within a finite system. This improbable, in principle negentropic possibility, allowed by thermodynamics, which would amount to an occasional and random return of the same, is hardly relevant for Stiegler. Not even the most improbable statistical probability could constitute an *avenir*. For, in the *avenir*, Stiegler attempts to confront the old Leibnizean question: 'Why is there something rather than nothing?'. Within a thermodynamic context this question assumes the form Poincaré gives it, by insisting that in a reality in which the laws of physics seem to demand the obliteration of life, 'life is an exception which it is necessary to explain'.[27]

The answer that Stiegler advances, is necessarily local and transient and *différance* is the way in which this answer is articulated. Negentropy constitutes accordingly 'a *différance*', insofar as it 'is a temporal deferral and a spatial differentiation of entropy'.[28] In order to make sense of this claim, entropy must be understood as the inherent *tendency* of a system, towards a state which is often designated as being of high entropy, but simply corresponds to a larger amount of possible microstates within a given macro-configuration, than one of lower entropy. In tandem, negentropy should be understood as a systemic arrangement that utilizes this tendency in order to achieve what Simondon calls metastability, that is, a state of functional flow of entropy, which by the same token amounts to a delay of its effects. In short then, entropy must be understood as *tendency* and negentropy not as the opposition or countering of this tendency, but rather as its 'appropriation', the participation in its unfolding in such a manner that something unforeseen emerges and lasts for a while.

It is precisely in order to avoid a misunderstanding of negentropy as a reduction or reversal of entropy that late Stiegler begins to employ more and more emphatically the term 'anti-entropy'. As one of his footnotes pauses to clarify: 'There can be no negative entropy: entropy irreversibly increases, whereas negative entropy would imply a reversibility that Carnot, Clausius and Boltzmann all rejected. [As mentioned, contemporary physics recognizes the theoretical possibility of this staggeringly improbable occurrence as a Boltzmann fluctuation.] This is why Bailly and Longo, like Wiener, refer to anti-entropy. Anti-entropy is what locally defers the irreversible increase of entropy, and it is as such that it is, in a strict sense, to the letter (a), a *différance*.'[29]

Perhaps however neither 'negentropy' nor 'anti-entropy' can altogether avoid the misunderstandings of a tragic agon of forces or of a dialectical synthesis of their operation. In order to do justice to the thrust of the meaning of negentropy or anti-entropy *thus* understood and to Stiegler's genial address of the thermodynamic question by means of *différance*, this reading proposes a conjecture for the need of a reformulation in which *différance* is not limited to the designation of negentropy as the difference and deferral of entropy,[30] but encompasses the becoming negentropic of entropy, which unfolds in parallel to the becoming entropic of negentropy.[31] This conjecture, in its preliminary outline, poses *différance* as a *question* within the future of the neganthropological project.

The question in sum can be articulated as follows: if the fundamental laws of physics work towards the eclipse of differentials (heat, pressure or other types of gradients) and if the negentropic organization of these differentials into metastable structures is only a means of intensifying the process of this eclipse, how can the locus of entropy be distinguished from that of negentropy? How can one utter one word instead of the other? How can any phenomenon be anything but (neg)entropic and precisely only *as* phenomenon either the one, or the other? Insofar as *every* negentropic structure embedded – parasitically – within a differential ends up producing *more* entropy than would have been produced without it, insofar as *every* negentropic structure becomes thus an instrument for the production of entropy and the spatio-temporal locality that serves as a *katēchon* of entropy serves also as its catalyst, both agonistics and dialectics falter.

A different configuration can however be sought in Stiegler's short breath, his double accusation of Heidegger: for failing to think thermodynamics in general, and more specifically, 'negative entropy, which, according to Schrödinger, is life'[32] and for failing to 'see that meditation, which forms the stakes of *Gelassenheit*, presupposes calculation, just as negentropy presupposes entropy.'[33] If breath itself were to serve as the simile of this double accusation in verso, Stiegler's postulate would read: oxygen *is* breath, just as carbon dioxide is the *presupposition* of breath. For indeed, far from a mere waste by-product of breath, carbon dioxide is the vehicle of breath, exchanged incessantly by the lungs with oxygen entering the bloodstream. Just as with negentropy, it is this partition of an exterior from an interior, this 'passing into the bloodstream', that makes life out of oxygen. Expectedly, a simile will only go so far – carbon dioxide is a compound distinct from oxygen, not a degradation of the latter, and is equally vital in photosynthesis as in breath, giving life twice over and linking together the broader carbon cycle. In contrast, entropy is a vehicle used a single time, the empty shell of a projectile, a vector of a unique trajectory. Negentropy rides this vector and becomes distinct only in the spatio-temporal partition it effects, in order to strengthen the becoming of entropy. In turn, only the becoming of entropy is able to bolster a negentropic structure. The negentropic becoming of entropy and the entropic becoming of negentropy are drawn in the *différance* of a continuous breath, the long *durée* of universal expiration.

To appreciate the implication of *différance* as the superposition of negentropy on entropy, the second question, that of 'translation', must be raised. For Stiegler, following Schrödinger, negentropy or anti-entropy is the answer that life gives to the second law of thermodynamics. Importantly, life does not 'solve' an equation of universal physics to achieve the maximum deceleration of entropy – minimum entropy would be incompatible with life. Rather, as it is apparent by now, life works *with* entropy, in a form of symbiosis or parasitism, which serves life's improbable origin and incalculable *telos*.

Then, out of life, a second moment emerges; in part as the unforeseeable future of life, in part as a solution to life's own toxicity. This is the moment of '*the passage from the organic to the organological*', the moment when tertiary retentions in the form of inscription and exosomatization establish the noetic soul and the 'human being' at large.[34] This organology is both a solution to the constant threats that life faces, as well as the origin of a new set of threats; in Stiegler's words this is a '*negentropic bifurcation*' which cannot but appear '*crazy or foolish*' since it is 'always capable of destroying the one who fabricates it, or the world in which it has been fabricated'.[35] In sum, life emerges as a second-order meta-solution to the inescapable imperative of the second law of thermodynamics, while technics is cast as the third-order meta-solution to the imperative of life, namely death.

The force of the neganthropological project lies precisely in the recognition and exploration of the possibility of these 'translations', these passages from the sub-atomic to the cell, and again, to *logos* and the algorithm. Among these 'bifurcations' the most decisive is for Stiegler the passage from 'vital' to 'noetic différance',[36] constituted by exosomatization, and constituting in turn a 'neganthropy'.[37] What poses itself as a future question, is the way in which such bifurcations or passages are to be understood and configured. At the same time, in order to enhance translatability, each level of analysis will have not only to deepen, but also to broaden: the thermodynamic plateau will have to open up to quantum and chaos theories in order to live up to the 'real' possibilities of the time that remains.

It is *différance* that vouches for the remaining time in the openness of the above questions. The metastabilization of forms within an entropic becoming and their translation from material to vital and noetic modalities is an operation always under negotiation. Inscriptions of highly divergent temporalities converge upon the spatio-temporal localization of each form and are themselves negotiated in the process of shaping it. This negotiation can be thematized and theorized as 'metabolics' – a task undertaken elsewhere.

The openness of the mutual becoming of (neg)entropy and its constant translation are sustained by the openness of the thermodynamic system in which (neg)entropic *différance* plays out. In fact, in a manner that Heidegger would term 'equi-primordial', it is the opening of (neg)entropic *différance* that sustains the openness of a thermodynamic system and establishes the *katēchon* that prevents the system from folding upon itself, from closing into a quick entropic death. Such openness however comes at the expense of further entropy; the ethico-political ramifications are unmissable.

Bernard Stiegler never flinched away from the task of thinking these ramifications, from pursuing these 'translations' and it is for this reason that he should be considered a philosopher of life. In a sense, for him the question *as well as* the answer is: life. Stiegler is not seeking a body without organs, yet he is also not seeking organs without a body. Even when organs detach themselves and set out on their own destiny beyond the body, to foster a general organology, Stiegler's demand is that they remain alive and that they foster life – 'exorganisms' must be invented that will enable 'new forms of locality' and sustain a 'neganthropy' for the time that remains.[38]

Life is thus for Stiegler the first bifurcation that confronts the 'arche-protention of the noetic soul' that it postpones and which constitutes it as a *Dasein*, namely death.[39] In the face of death, knowledge 'is a care that technical life takes of itself'.[40] The Platonic injunction of philosophy as a study of death finds thus in Stiegler its most living expression, as one begins to recollect its memories of the future.

Notes

1 Daniel Ross, 'Introduction', in Bernard Stiegler, *The Neganthropocene*, trans. Daniel Ross, London: Open Humanities Press, 2018, 7–32, 22.
2 Søren Kierkegaard, *Repetition*, trans. Howard V. Hong & Edna H. Hong, Princeton: Princeton University Press, 1983, 131.
3 Bernard Stiegler, 'What Is Called Caring? Thinking beyond the Anthropocene', in Stiegler, *The Neganthropocene*, 188–270, 202.
4 Stiegler, 'What Is Called Caring? Thinking beyond the Anthropocene', 209.
5 Bernard Stiegler, *Technics and Time, 1: The Fault of Epimetheus*, trans. George Collins & Richard Beardsworth, Stanford: Stanford University Press, 1998, 38.
6 Stiegler, *Technics and Time, 1*, 54.
7 Stiegler, *Technics and Time, 1*, 61.
8 Stiegler, *Technics and Time, 1*, 68.
9 Stiegler, *Technics and Time, 1*, 69.

10 Bernard Stiegler, *Technics and Time, 2: Disorientation*, trans. Stephen Barker, Stanford: Stanford University Press, 2009, 98.
11 Stiegler, *Technics and Time, 2*, 143.
12 Bernard Stiegler, *Technics and Time, 3: Cinematic Time and the Question of Malaise*, trans. Stephen Barker, Stanford: Stanford University Press, 2011, 74.
13 Stiegler, *Technics and Time, 3*, 76.
14 Stiegler, *Technics and Time, 3*, 134.
15 Stiegler, *Technics and Time, 3*, 171.
16 Stiegler, *Technics and Time, 3*, 212.
17 Stiegler, *Technics and Time, 3*, 171.
18 Stiegler thanks Benoît Dillet, for reminding him Claude Lévi-Strauss's text in the General Organology conference he and the Nootechnics group organized at the University of Kent, in November 2014. Bernard Stiegler, *Automatic Society, Volume 1: The Future of Work*, trans. Daniel Ross, Cambridge: Polity Press, 2016, 320.
19 Stiegler, *Automatic Society, Vol. 1*, 242; Stiegler, 'Escaping the Anthropocene', in Stiegler, *The Neganthropocene*, 51–63, 56.
20 Claude Lévi-Strauss, *Tristes Tropiques*, trans. John Russell, New York: Criterion Books, 397.
21 Lévi-Strauss, *Tristes Tropiques*, 397.
22 Stiegler, *Automatic Society, Vol. 1*, 243.
23 Bernard Stiegler, 'Fives Theses after Schmitt and Bratton', in Stiegler, *The Neganthropocene*, 129–38, 131.
24 Erwin Schrödinger, *What Is Life?*, Cambridge: Cambridge University Press, 2013, 7.
25 Schrödinger, *What Is Life?*, 85. Emphasis added.
26 Bernard Stiegler, *The Age of Disruption: Technology and Madness in Computational Capitalism*, followed by *A Conversation about Christianity with Alain Jugnon, Jean-Luc Nancy and Bernard Stiegler*, trans. Daniel Ross, Cambridge: Polity, 2019, 19.
27 Nicholas Georgescu-Roegen, *The Entropy Law and the Economic Process*, Cambridge, MA: Harvard University Press, 1971, 189.
28 Bernard Stiegler, *The Nanjing Lectures 2016–2019*, trans. Daniel Ross, London: Open Humanities Press, 2020, 303.
29 Stiegler, *The Neganthropocene*, 304 n. 416. Elsewhere, Stiegler writes: '[T]here can be no such thing as negative entropy to the extent that entropy, as the dissipation of energy, is irreversible. In the strict sense, "negative entropy" would mean going backwards in time: only then would entropy be reversible. There may be, however, a *local limitation*, a *temporary deferral* of the entropic "penchant" of inert matter, through a vital process of the accumulation of energy' (Stiegler, *The Nanjing Lectures*, 305).
30 Bernard Stiegler, 'Welcome to the Anthropocene: Text for an Encounter between Bernard Stiegler and Peter Sloterdijk', in Stiegler, *The Neganthropocene*, 103–14, 103.
31 The issue is hinted at here: 'the trace (or the archi-trace [...] of *différance*, and that is *différance*)'. Stiegler, *The Nanjing Lectures*, 303. The trace, in this instance the formations of negentropy must be taken as a trace of *différance*, not *différance* itself.
32 Stiegler, 'Welcome to the Anthropocene', 105.
33 Stiegler, 'Welcome to the Anthropocene', 105.
34 Stiegler, 'Escaping the Anthropocene', 62.
35 Stiegler, *The Age of Disruption*, 102.
36 Stiegler, 'Welcome to the Anthropocene', 103.

37　Bernard Stiegler, 'Passages to the Act: Dialogical Interactions and Short-Circuits in Interactivity', in Stiegler, *The Neganthropocene*, 92–102, 92.
38　Bernard Stiegler, 'Governing Towards the Neganthropocene', in Stiegler, *The Neganthropocene*, 115–28, 127–8.
39　Stiegler, 'Passages to the Act, Dialogical Interactions and Short-Circuits in Interactivity', 92.
40　Bernard Stiegler, 'Capitalism as Epistēmē and Entropocene', in Stiegler, *The Neganthropocene*, 139–51, 140.

Chapter 9

WHERE THERE IS NO WORLD AND NO EPOCH: BERNARD STIEGLER'S THINKING OF THE ENTROPOCENE

Erich Hörl

Introduction: Epochal thinking

A long quotation from the chapter 'On a Change of Epoch: The Exigency of Return' in Maurice Blanchot's *The Infinite Conversation* marks the threshold of *Technics and Time*. The passage – a dialog not about a new period in history, but about a caesura, a turning, a step beyond, operated by modern technology, and about the extreme danger that characterizes this transition to what appears to be absolutely other – stands emblematically above the general introduction to Stiegler's multi-volume project. It is the cue for an entire oeuvre subsequently pursued in numerous ramifications – an oeuvre that in a sense will have been a vast interpretation of Blanchot's reflection on the change of epoch. *Incipit* as follows:

– Will you allow as a certainty that we are at a turning point?

– If it is a certainty, then it is not a turning. The fact of our belonging to this moment at which a change of epoch, if there is one, is being accomplished also takes hold of the certain knowledge that would want to determine it, making both certainty and uncertainty inappropriate. Never are we less able to outline ourselves than at such a moment, and the discrete force of the turning point lies first in this.[1]

This text was previously presented in a first, shorter version at the Leiden Center for Continental Philosophy's conference *Memory of the Future: Thinking with Bernard Stiegler* and, in a longer version, in the Seminar of Aesthetics at the University of Oslo. I would like to thank Susanna Lindberg (Leiden) as well as Ina Blom, Ursula Münster and Eivind Røssaak (Oslo) for inviting me; and I would like to thank all participants for the discussions and some valuable suggestions.

Since Blanchot's reflections from 1960, a new computational condition has emerged. In this moment of turning, the difficult task of 'outlining ourselves' has become an almost impossible challenge. The 'we' that is to be outlined and reconstituted is, as such, falling apart, undone by all kinds of processes of disindividuation Blanchot could not have dreamed of. This disintegration of the 'we' is an essential moment of the turning. Yet Stiegler accepts the challenge, accepts it as the task that defines his life as a philosopher. The *problem* of the 'change of epoch' and the development of this problem in, or as, the *question* of epochality – which raises the problem in an absolutely original way from the perspective of technology and goes as far as to reconceive the techno-logical meaning of epochality, in which a 'we' is composed and consists in the cooperation of a *who* and a *what*[2] – they mark and situate Stiegler's thinking from the outset and define its structure and becoming throughout all the phases it goes through – from the technological to the pharmacological-organological and finally to the neganthropological.[3]

'[T]hat we are at the end of one discourse and, passing to another, we continue out of convenience to express ourselves in an old, unsuitable language. That is the greatest danger. It is even the only one'. In the wake of 'the seizure of the constitutive forces of matter' and 'the supremacy of the machinelike play of these forces', Blanchot writes, 'the terminology proper to *historical* time' – above all concepts such as 'freedom, choice, person, consciousness, truth, [and] originality' – can no longer do justice to 'the event we are encountering' that henceforth 'bears an elementary character: that of [...] impersonal powers'. In his view, the Romantic interpretations of technocracy proffered by the likes of Ernst Jünger or Teilhard de Chardin betray the insight '[t]hat, hidden in what is called modern technology, there is a force that will dominate and determine all of man's relations with what is. We dispose of this force at the same time as it disposes of us, but we are ignorant of its meaning; we do not fully understand it'. The danger, then, lies 'first of all in our refusal to see the change of epoch and to consider the sense of this turning'.[4] The remarkable conceptual richness of Stiegler's work, the peculiar conceptual politics that characterizes it, his enormous creative capacity for new descriptions as he struggles against the loss of language and concepts brought on by modern technology by engaging with numerous, not just philosophical but anthropological-ethnological, physical, economic, biological, neurophysiological, computer science discourses: all this is prompted by and founded in the danger that according to Blanchot confronts us at the brink of historical ages. It is undoubtedly an attempt to move from an old terminology inappropriate to the epochal transformation to be witnessed, to a comprehensive reconceptualization of the technological condition as such, a new description that can do justice to the epochal sense of the 'we' – the sense of the 'we' that always writes an epoch, that constitutes epochality – an account that may well allow us consistently to think this sense for the first time.

Stiegler focuses Blanchot's problematization and elaborates it as the relation between technics and time. This not only brings into view the change of epoch taking place in the course of twentieth-century computerization and cybernetization within the *longue durée* of the history of grammatization, where history is opened up as such by writing, where it develops *as* writing and thus always already as a techno-logical history. The change of epoch on the brink of historical times, which is at the same time a threat to history posed by its inner outside, technology – and we will see how Stiegler is haunted especially by the definition of this brink, which he ultimately situates at and describes as 'the disruption', thereby naming our contemporary impossible epochality at and as the extremity of the Entropocene – sheds a whole new light on the entire problem of history as such. In the end, Stiegler is concerned with completely redefining the logic of epochality itself, brought out like never before by the change of epoch at the zenith of history, with reconstructing it in its essential technologicity, and with discovering the sense of this very turn precisely in this and as this redefinition: History appears as a time that unfolds as an always already belated, delayed improbable response to a technological suspension, as a differing time, whose cliff or abyss would be a humanity – or, as Stiegler following Jacques Derrida says, a *Geschlecht* – outside of the time of the delaying-delayed that would open up a future and thus a history, the *Geschlecht* of a 'time without time', a *Geschlecht* in the timeless and the epochless, in the collapse of the ability to respond, incapable of differing transformation.[5] It is a sense of sense, trans-formative all the way down, taking shape in the epochal turn as its challenge, that Stiegler's work in ever new attempts seeks to conceptualize and to situate in the history of sense. In the moment of its sudden interruption, his oeuvre leaves us with the urgent task of transformation – the task of careful thinking. Stiegler's thinking is epochal, in the most emphatic sense of the term. That is what I hope to show in what follows.

A thinking of suspension: Exploring the logic of epochs

The thinking of epochal change that Stiegler proposes within the scope of his thinking of suspension begins with an absolutely subversive rearticulation of the problem of *epochē* that gives rise to his unique philosophical, diagnostic and political project. For him, *epochē* is no longer, as it is for Edmund Husserl, a *methodological* principle of 'parenthesizing' or 'exclusion' of the natural world (of what is simply there, present at hand), no longer an operation of a transcendental ego that achieves the transition from the natural to the apex, so far, of the theoretical attitude.[6] Prior to every *epochē* and as its originary supplement, we might say, there is something Stiegler calls 'objective *epokhē*'. This objective *epochē* – prompted by thrusts of exteriorization on the part of technical objects or, in the terms he borrows from Alfred Lotka in the late, neganthropological phase of his work, by exosomatization – turns out to be the heretofore unthought condition of the epoch, the background of all epochality. This reversal of the problem of *epochē*, which shifts from subjective *epochē* as a methodology to objective *epochē* as a central moment of history, initially grounds his pharmacological-organological work, which, starting in *The Fault of Epimetheus*, the first volume of *Technics and Time*, exhaustively presents the development and ramification of this reversal. And the transition to the later neganthropological phase will take up the *epochē* problem once more, refine it conceptually and make it the decisive moment in the historical definition of our present as being without epoch and world. The central concept in this vast effort of rearticulation, which both technicizes and politicizes phenomenology, is 'epochal redoubling'.

At the very latest, at the beginning of *Disorientation*, the second volume of *Technics and Time*, the reflection on epochal redoubling comes to perform an especially important conceptual function. In those pages, it becomes the essential moment that lifts the problem of the change of epoch (which defines the entire enterprise from the outset) onto a new conceptual level that will shape Stiegler's thinking all the way to his philosophical-political strategies. The decisive passage reads as follows:

> In *The Fault of Epimetheus*, I demonstrated that the reification of a technical propensity or body of propensities, leading to an altered technical system, suspends the behavioural programming through which a society is united, and which is a form of objective *epokhē* the social body initially tends to resist. An adjustment then takes place in which an epochal intensification [*redoublement épokhal*] occurs; this adjustment is the *epokhē*'s key accomplishment, in which the *who* appropriates the effectivity of this suspension (i.e., of programmatic indetermination) for itself. Technical development is a violent disruption of extant programs that through redoubling give birth to a new programmatics; this new programmatics is a process of psychic and collective individuation.
>
> Contemporary disorientation is the experience of an incapacity to achieve epochal redoubling.[7]

In fact, though, Stiegler presents the movement of epochal redoubling in a very early text on Charlie Parker and the gramophone that considerably predates the publication of *Technics and Time* – and it seems that, inversely, the entire subsequent development of the problem of technics and time seeks to explicate the movement first sketched here. Its discovery marks Stiegler's most moving, most consequential thought:

> The endurance of the improbable – that is to say: the suspension of pro-grams, clichés, attitudes, gestures, pre-judged words and actions, badly repeated stereotypes – stems from what philosophy calls *epokhē*, from common Greek. *Epokhē* means at once an interruption, the suspension of judgement, and the state of doubt: the point in the sky where a star seems to stop, a period of time, an epoch, an era.[8]

Having indicated the philosophical use of the word, especially Husserl's use of *epochē* as 'parenthesizing' and Martin Heidegger's reformulation of the question of *epochē* in terms of epochs of Being and of the withdrawal of Being, Stiegler emphatically returns to his own use of the term:

Epokhē, as I have come to use the word here, is first the very actuality of historical time, history in action: this epoch within the succession of epochs. That is how we understand the primary sense of the word today. But, to return to philosophical tradition, and against it, my aim is to argue that, just as Charlie Parker made history with his saxophone *and* another instrument (the phonograph), *epokhē* is always *double* and always supposes an *epokhal techno-logical ground*. *Tekhnē* suspends an epoch from *tekhnē*; *tekhnē* makes *epokhē*, and, in this suspension, there is an improbable response, a linkage, a making of time: it is *epokhē* that makes an epoch.⁹

And, more forcefully still: 'What *tekhnē* requires is a "response", a linkage that is im-probable, which it cannot as such give.'¹⁰ Technics is epochal in the strongest sense possible: it interrupts an epoch and thereby, precisely, in this interruption and as a programmatic reply this interruption demands, it inaugurates an epoch.¹¹

The fact that this early publication already contains in nuce this key thought points to the constitutive role the *epochē* problem plays in the genesis of Stiegler's philosophy – of his problematic, to speak with Althusser – as a whole. It even ties in with how he became a philosopher, at least in his account some twenty-five years later: even before any reading of Husserl, he says, he 'discovered what one calls in philosophy the phenomenological *epokhē* – the suspension of the world' in 'a perfectly singular experience' of the withdrawal of the world that he had during his incarceration in Saint-Michel near Toulouse. With all the apodicticity that characterizes his later work, he was then 'able to pass from the empirical practice of the *epokhē*, of the suspension, to a practice that was reasoned, methodical'.¹² This passage was to fundamentally revise the conception of *epochē* (henceforth articulated, as we have seen, as epochal redoubling), deconstruct the meaning of *epochē* in a countermove to the metaphysical inscription that in Husserl is the decisive act of transcendental subjectivity and decipher it as a historical momentum of techno-logical assemblages.

At its theoretical core, *The Fault of Epimetheus*, which presents a first great elaboration of the problem of the epoch, initially by interlacing Bertrand Gille's, Gilbert Simondon's and André Leroi-Gourhan's reflections on technological evolution, is already marked by a conspicuous semantics of doubling – the legendary 'doubling-up' of the lonely figure of Prometheus by that of his brother Epimetheus, and the concomitant 'double fault' at the origin, which, in an 'originary doubling-up', makes Prometheus's fault appear as a doubling-up of Epimetheus's fault, as 'two moves [*coups*]' and an 'originary duplicity' that characterise exteriority as the 'originary incompleteness of technical being'.¹³ Shining through this semantics of doubling is the signature of a new thinking of epochality, which re-examines the entire problem of epochality, the peculiar movement of a techno-logical differing, as the central moment of historicality. According to Stiegler, it is the techno-logical differing that produces differences. To answer the question 'What is the epochal?' that structures the final section of *The Fault of Epimetheus* – the question concerning how an epoch of historicality can open (up to us) at all – Stiegler mobilizes the figure of epochal redoubling that technologically repeals the established forms of a previously programmatic tradition and at the same time programmes a new enduring of the past, of anticipation and of the present.[14]

Yet it is only in *Disorientation* that the entire conception comes to hinge on the basic problem of epochality.[15] Stiegler transforms the *problem* of the change of epoch into the *question* of the change of epoch, which allows for thinking this change – *penser* – but thereby also for healing it and caring for it – *panser*.[16] From the outset, epochal redoubling serves as the decisive formula for unlocking the sense of epochal change as a trans-formative sense that lays out the movement of transformation in precise detail as *the* basic historical movement. At the same time, it develops the new perspective of problematization introduced by objective *epochē* in the direction of a vast diagnosis of the epoch yet to be given and of a downright politics of the *epochē*:

> If *tekhnē* suspends the programs in force, then knowledge also returns to suspend all stable effects, *tekhnē*'s 'repercussions', by redoubling them. This is *epokhal redoubling*. [...] Linear and phonological writing is a programmatic *epokhē* suspending all forms of a heritage that is itself programmatic but as such does not appear to be, and which, in suspension, pro-grams an other vestige of the past, of anticipation, and consequently of a present conceived as presence. Which idea of today, then, would (improbably) program the *epokhal redoubling* of *différant* analogic, numeric, and biologic identities, thus throwing into crisis the presence of which 'today' consists?[17]

In the notion of a crisis of the present, this passage not only names a threat to, if not destruction of the capacity for epochal redoubling – a fundamental disorientation around which the totality of Stiegler's critical and conceptual project and his politics especially, as we will see, will henceforth revolve. And that, precisely, is what he elaborates, single-mindedly, in the last stage of his oeuvre, beginning with the transformation of his pharmacological-organological thinking into a neganthropological critique of the Entropocene under the heading of being-in-disruption. Inversely, we can see in the definition of being-in-disruption the most consequential conceptualization of the initial intuition of the problem of the epoch. Moreover, in the passage just quoted, the entire enterprise of deconstructing the metaphysics of presence – a tradition Stiegler himself is part of – as such proves to be already an essential part of an epistemic reply to a techno-logical shock, a reply that forcefully brings out an entire epoch's fixation on the present, interrupted by this shock. As it renders the project of deconstruction legible, Stiegler's thinking of suspension radically develops deconstruction in the direction of what remains unthought in it: the unthought of technics.[18] It thereby specifies 'deconstruction' – if it exists as such – as to its historical condition: deconstruction is prompted and made possible in the first place in and by the entry into the age of cybernetics and thus of the computational *Gestell* that puts an end to the epoch of phonological writing, including the sense of historicality this epoch has shaped. The thinking of suspension captures these historical stakes of deconstruction, whose full significance is thereby presented for the first time.

Probably the most precise articulation of the concept of the 'doubly epokhal redoubling' can be found in *The Age of Disruption* – the most precise because there, Stiegler thinks it in terms of its specific temporality, namely as a temporal redoubling or a composition of two times, and spells out the implications of this temporal composition for conceptualizing historicality and epochality. Here is the passage, which brings out the temporality of the formation of epochs and is of such importance for everything that follows:

> When a technical system engenders a new epoch, the emergence of new forms of thinking is translated into religious, spiritual, artistic, scientific and political movements, manners and styles, new institutions and new social organizations, changes in education, in law, in forms of power, and, of course, changes in the very foundations of knowledge – whether this is conceptual knowledge or work-knowledge [*savoir-faire*] or life-knowledge [*savoir-vivre*]. But this happens only in a second stage [*un second temps*], that is, *after* the techno-logical *epokhē* has taken place.
>
> This is why an epoch always occurs through a doubly epokhal redoubling:
>
> – *double* because it always occurs in *two stages* [*en deux temps*] – on the one hand, the technological *epokhē*, on the other hand, the *epokhē* of knowledge as forms of life and thought, that is, the constitution of a new transindividuation (characteristic of a particular time and place);
> – *redoubling* because, starting from the *already there* forms of technics and time that are constituted as this or that established epoch, a new technical reality and a new historical reality (or, more precisely, historial – *geschichtlich*) redoubles and through that relegates to the past that which has engendered it, which seems, therefore, precisely to be *the past*;
> – *epokhal* because it is only as an *interruption* inaugurating a *recommencement* and a *new current present* that this double redoubling occurs, eventually by firmly establishing itself as what we call, precisely, an *epoch*.[19]

Even the very function of reason, Stiegler suggests in one of his key late texts, hinges on the doubly epochal redoubling: '"Rational", here, does not mean logical, referring to apodictic truth as canon and so on (the logical and the apodictic constituting a specific configuration of the exosomatization of noetic functions and faculties), but *everything that assumes the function of reason as the capacity to effect a noetic bifurcation after a doubly epokhal redoubling*.'[20] Not only that: what once enjoyed veneration under the heading of the History of Being – a title that focuses Heidegger's problematization of epochality as a succession of epochs of the forgetting of Being, without it ever becoming clear where these are written from – is driven by it as well: 'The "history of being" is a succession of such crisis generated by the tensions provoked by the succession of doubly epokhal redoublings in which the process of exosomatization and its acceleration consists.'[21]

What we find concentrated in the formula of epochal redoubling is nothing more and nothing less than the core problem of technics and time as such, whose diagnostic force and reach clearly exceed those of being and time.

Being in the disruption: In the worldlessness and epochlessness of the entropocene

The full stakes of Stiegler's thinking of suspension are apparent where it develops its diagnostic power to define the present epoch – a present epoch, as we will see in a moment, that marks the limit of the logic of epochs as such, its nihilistic ending, as non-epoch or without epoch. This diagnosis is organized by the definition of the doubly epochal redoubling, which it forcefully reasserts and elaborates in broader and more profound terms. It is the doubly epochal redoubling – and this, precisely, is what Stiegler designates as the key characteristic of our 'today' – that is pierced, damaged, put into question as such. This displaces the very logic of epochs, which is founded on that redoubling and supports the movement of historicality as a whole.[22] The entire sense of Stiegler's rethinking of the epoch or of the change of epoch, which ultimately defines the heritage of his prodigious elaboration of the question of technics and time, culminates in this diagnosis, a diagnosis that ultimately appears as the horizon of the question of technics and time as such, and as that which is at stake in and compels the elaboration of that question as *our* question. The outlining of what I call the un-time of epochlessness – the 'un-', once again, of a double lack: a lack of time as of epoch – brings out the core of the entire enterprise, endows it with its urgency and reveals its historical signature as the collapse, which it allows us to think, both of historicality and temporalization.

Now, what does epochlessness mean, precisely? And why can it be characterized as an un-time? If the improbable, indeterminate answer to the suspension (in which the belated redoubling of the techno-logical *epochē* takes place) is not given and the change of epochs does not succeed; if, therefore, the technological interruption – of the programme, of tradition, of inheritance, of the systems of care that together make up an epochality – is absolute, that is, if there is disruption (which is what the disruptionists of the great platform industries dream of, what they constantly seek to implement); and if all epochality threatens to burst, all existence starts to dissolve in an epochal void bereft of responses and responsibilities, and all consistency begins to get lost, then we find ourselves in the infinite deformation and interruption of epochlessness – such as it characterizes our *being in the disruption* Stiegler describes as our contemporary condition of being. Not every interruption that inaugurates an epoch is thus disruptive. There is disruption only when the interruption cannot be redoubled and an absolute void of thinking is permitted. '*The* disruption' is thus a *historical* definition through and through. The detailed analysis of the doubly epochal redoubling in *The Age of*

Disruption refers to the two temporalities that interlock here – technological time on the one hand and the time of knowledge, of forms of life and of thinking on the other. In disruption, the second time of epochal redoubling is missing.

What is missing, at the core, is the time of questioning that is able to open up the horizon of the unexpected in the technological interruption. Where Heidegger (as Derrida has shown) privileges questioning in thinking, where he again and again equates thinking and questioning – declaring questioning to be the 'piety of thinking' – but does not question the possibility of questioning any further, there Stiegler discovers this very possibility as such:

> But *Dasein* can question only because it is itself put into question. And what puts it into question is the *pharmakon*. The *pharmakon*, as a technical (exosomatic) upheaval, is what puts into question the one who questions, which is to say the very possibility of questioning. *Dasein*, the privileged being, questions only inasmuch it is put into question by that what precedes it and at the same time exceeds it beyond all questions, thereby forming what Bergson called an obligation.[23]

The possibility of questioning is directly connected to the basic structure of the doubly epochal redoubling, which grounds all experience of the question. The refoundation of the question thus given leads to a redefinition of thinking as careful thinking because thinking, coming out of the questioning by technics is to give a differantial turn to this questioning, thereby exercise a 'therapeutic activity' and heal the 'wound' – the wound of (technological) suspension.[24] As Stiegler points out, '[t]o think would therefore be to take care, to care for, which is also to say, to act, to make (the) *différance*: it would always be to think the wound.'[25] This means, inversely, that in the disruption, this very possibility is under threat to a previously unknown extent – Stiegler even speaks of the production of *différance* collapsing[26] – that a disastrous rule of the unquestioning begins. Under these conditions of increasing questionlessness, there is, in particular, no more time of transindividuation, a differantial time, that is, in which, through the diachronous co-individuation of the 'I', the synchronizing trans-individuation of a 'we' could take place that in turn transforms the 'I's. All that is left is the non-time of a loss of individuation, of disindividuation, in which the 'we', too, loses its consistency, in which no consisting and no existing succeeds any more: 'In the midst of disruption, the second stage of the doubly epokhal redoubling fails to occur; there is no transindividuation.'[27]

This is the problem at the centre of Stiegler's last work, which undertakes to elaborate the question of the completed nihilism of the computational *Gestell* – an undertaking he places, quite logically, under the heading of an 'absence of epoch'.[28] The loss of world that Heidegger in his thinking of the *Gestell* problematized as the horizon of technicized modernity and its boisterous effort at world-formation, the loss Gérard Granel, one of Stiegler's great teachers, then reconstructed as a great process of unworlding resulting from the conjunction of infinitization operated by technics and capital, this deworlding, in Stiegler, becomes total, the final consequence of epochal absence in the disruption. What is without epoch is without world; in other words: Stiegler's thinking of epochlessness appears as an interpretation and conceptualization of the problem of unworldliness.[29] If for Stiegler, 'world' is the sense-assemblage of a consisting based on an always already externalized, essentially exosomatic *différance*, then the total destitution of epochlessness – if the collapse of the movement of differing itself is even possible – is the culmination of the deworlding and unworlding at the high point of modernity.

Stiegler understands this absence – we might even say that the time of presence (*Zeit des Anwesens*) is replaced by an un-time of absence (*Un-Zeit des Abwesens*) – above all as an 'absence of shared and cultivated protentions'[30] that are replaced by the automatic protentions of algorithmic governmentality and their destruction of horizons of expectation, of futurality, and of time generally. The Time-Form that characterizes the basic historical-ontological constitution of disruption – what Antoinette Rouvroy and Brian Massumi call 'pre-emption', Mark Hansen 'feed-forward', Stiegler 'tertiary protention', to give some examples of the Time-Form emerging in the computational *Gestell*[31] – the Time-Form of environmentality, ultimately proves to be an un-form of detemporalization that casts a fundamental absence over the entire epoch, which thereby completely turns into epochlessness. The foregrounding of tertiary protentions by digitality and their industrialization and automation by algorithmic governmentality dislodge the epoch-making character of the retentional apparatuses operating in historical ages. According to Stiegler, 'the accumulation of collective retentions and protentions', which make up a retentional apparatus, 'forms epochs'. Epochs are constituted by the division of retentions and protentions; the computational tertiary protentions are disruptive because they assume control of this division.[32] Finally, all time disappears in eventlessness: 'In the absence of epoch, where nothing can be acknowledged, nothing can happen [*Dans l'absence d'époque où rien ne peut être avéré, rien ne peut advenir*].'[33] Concentrated in the concept of the absence of epoch, no doubt, we find the immense philosophical diagnosis of the coming un-time – strictly speaking an un-time of the non-event – begun in *Technics and Time, 1* at the latest but reaching its destination only now. The generalized denoetization and total proletarization that undermine the process of doubly epochal redoubling result in the 'nonindividuation that is the absence of epoch'.[34] Absence of epoch, then, is the name of the devastation wrought by the absolute disruption, which perfects the technological nihilism of being in the disruption. We live in the impossible epoch of the absence of epoch, in a missing epoch in which no consisting succeeds that might otherwise form an epoch and in which the 'we' of an epoch might appear. Is this the general form of our today, which, according to Stiegler, all thinking worthy of the name must aim to define?[35]

This definition sheds new light also and especially on the problem of care; it might even make it the central theme of our epoch (if we can still put it like that), as we saw above in the context of Stiegler's rethinking of the question or of what is without question: 'Such an absence of epoch is affected more than ever *by the problem and by the question of care.*³⁶ The question of care in Stiegler proves quite simply to be *the* question of epochlessness that befalls us today and thereby is *our* question – even if this 'we', whose problem and question it is, is increasingly absent, as a consequence of the general carelessness. Strictly speaking, it is only thanks to Stiegler that the problem becomes the *question* concerning care, which directly follows the question concerning technics and even becomes its core content. In the disruption, the problem of care powerfully comes out as such and becomes recognizable as a *question* in its direct link with the question of technics; indeed, it proves to be at the core of the question of technics. The structure of care as such is folded into the basic structure of doubly epochal redoubling: it arises, precisely, from its temporality, which interlaces having-been, future and present in a peculiar way, from the fundamental questioning.³⁷ As caring for the *pharmaka* of technics, care is always care for *différance*, which, however, in the possible transition to 'an age of *indifférance*', proves to be our central challenge as well as the task of careful thinking.³⁸

It is this epoch-, time- and carelessness of disruption that Stiegler finally addresses as the last stage not of the Anthropocene but of the Entropocene:

> the Anthropocene, then, is what is best characterized as a liquidation of localities and a generalised and planetary increase of thermodynamic entropy as an increased dissipation of energy; of biological entropy as destruction of biodiversity; and of informational entropy as destruction of noodiversity.
> Characterised this way, *the Anthropocene can just as well be called Entropocene.*³⁹

Especially the loss of noodiversity and the general denoetization in the digital disruption, which according to Stiegler implement a structural carelessness, put our responsibility – understood in the sense of the doubly epochal redoubling as 'our ability to *respond* to the challenge of being put into question'⁴⁰ – at stake in its entirety, and with it, our capacity for transforming becoming into future, entropy into negentropy. Conversely, the central challenge of the Entropocene is the search for the '*possibility of a new type of doubly epokhal redoubling in the absence of epoch*' – what Stiegler calls '*neganthropological différance*'.⁴¹ At issue are improbable responses to the disruption that bring about a new great transformation – this, precisely, is why our position in the history of sense might be that of a transformative sense – that usher us into a 'new epoch', the 'curative, care-ful epoch' of the Neganthropocene.⁴²

Stiegler's general organology – with its three interlocking perspectives of psychical, technical and collective individuation – is an attempt to think in the entropocenic absence of epoch, to think nonetheless and despite everything, and in so doing – this is decisive – to re-evaluate the total nihilism of the disruption.⁴³ This makes organology the site where a new politics crystallizes: in the end, the thinking of suspension, whose diagnosis undoubtedly culminates in the interpretation of the Entropocene, proves to be a broadly conceived politics of the *epochē*. Stiegler's version of a techno-phenomenology of the spirit depicts our age of the computational *Gestell* after the end of history as an un-time of nonknowledge in which such a politics becomes above all a demand of thinking itself, because thinking [*penser*] means healing [*panser*]. In the digital disruption, thinking must care like never before for the responsible restoration of epochal (noetic) forms that allow for life on par with the technological condition. This thinking, which simply takes over the long-standing but obscured task of thinking, is healing. That is what the concept of careful thinking entails.[44]

At this late point, in *Qu'appelle-t-on panser? 1* Stiegler once more turns to Heidegger, and in particular to the Heidegger who in the 1951–2 lecture course *What Is Called Thinking?*[45] is struggling with the change of epoch Blanchot is writing about. Heidegger's new way of asking the question what thinking means can be read as an early attempt at a response to the techno-logical disruption that, at the time under the heading of cybernetics, was at least preparing itself and, de facto, inaugurated being in the disruption.[46] Stiegler's rereading – which also picks up on the great translation of Heidegger's lecture course by Granel – this rereading practically imposes itself at the beginning of the twenty-first century, when the *Gestell*, as computational *Gestell*, begins to reach its apex. Thinking means healing – this marks the core of the programmatic reprise of working through the task of thinking that, for Stiegler, Heidegger did not push far enough.

At bottom, though, two repetitions intersect here. In a peculiar way, the *a* that in Stiegler's reinterpretation of *penser* as *panser* replaces the *e* and thereby indicates a historical point in which the question of thinking as healing and the problem of epochlessness appear so radically, repeats Derrida's operation. Derrida famously replaces the *e* in *différence* with an *a* to mark the *différance* that, precisely, does not mean difference alone but the process of differing that simultaneously differs and defers, that can never be stabilized for good and that thereby temporalizes and spatializes, that figures the writing of life itself. Now, in Derrida already, this entire operation ultimately concerns thinking *our* epoch, which he himself, at the beginning of *Grammatology* (1967), inscribes in a movement of technology and which he associates with the emergence of cybernetics and the implementation of a new, a mathematical writing and thus with a far-reaching 'new mutation in the history of writing, in history as writing':[47] 'I would say, first off, that *différance*, which is neither a word nor a concept, strategically seemed to me the most proper one to think, if not to master [...] what is most irreducible about our "epoch".' Derrida, to be sure, already notes that we 'could no longer even call this an "epoch"' in the strict sense.[48] Stiegler captures this point in the precise terms of the epoch of the absence of epoch, of a missing epoch, and he articulates its pharmacological condition – and here, in this absence, this missing, he lets a new sense of thinking emerge: thinking as healing. The transition from a thinking of *différance* to thinking as *panser* that cares for and makes (the) *différance*, brings the problem of care as such to the fore and thereby inscribes deconstruction once more, and more forcefully, in its time – the time of disruption that reveals the fundamental connection of technics and care. Part of the heritage, if not *the* heritage of deconstruction, is to have outlined a caring thinking, as Stiegler has done.

Conclusion: *The melancholiform aspect in the ending of historical ages*

There is no doubt that the 'blinded lucidity' which distinguishes Stiegler's thinking of suspension made him suffer. It got to him. And he was aware that his entire diagnosis of automatic nihilism in the disruption – thought, on the basis of the economic-technological disruption in the digital transformation, of the innovations that interrupt all previous structures, as the subjection of society under models that destroy its structures and lead to total disindividuation – that the diagnosis, whose relentless elaboration drives his thinking and its peculiar sound, was an affront for his contemporaries and fell prey to their denial and rejection. He says that in the second decade of the twenty-first century, we found ourselves

> in the permanent and universal state of emergency of what appears to us doomed to become unliveable. We all have a sense of this *state of affairs*. But most of the time we deny it because it is *unbearable* – most of the time, *except when we cannot* but notice its *immediate, disastrous, and massive* effects in the everyday of our existence. Then, we are overwhelmed. Let's call these moments of blinded lucidity – in which negation and denial become *impossible* even as they *dominate*, thus provoking an immense melancholiform suffering this author knows only all too well – *negative recurrences* [*intermittences negatives*]. What is to be *done* – given *that we are* confronted with this more or less hysterical, melancholy, cyclothymic, 'bipolar' […] *intermittent negativity* – […] what is to be done such that, through the curative effects of practicing *always more than human*, if not 'superhuman' quasi-causality, these lucidly blind moments turn into moments of *positive recurrences*?[49]

A strange, a last 'we' crops up here that seems to appear once more in moments of negative intermittence, in the absolute danger of decay, already melancholiform. Once more we find here the problem of turning. This is the core movement of sense that Stiegler's techno-phenomenology brings out, a sense I would like, for that reason, to characterize as trans-formative. If sense, according to Stiegler, always has a transformative moment, if sense arises as a response, as the programmatic turning of intermittences that contest it, and if this movement of turning always endows with sense, then sense is being radically exposed as trans-formative sense only in the disruption that threatens completely to destroy this sense and to collapse into senselessness. The central endeavour, which Stiegler, taking up Derrida, Nietzsche, Heidegger, Canguilhem and Guattari, unerringly pursues over all these years, the endeavour to conceive thinking as healing and thereby pharmacologically to redevelop the problem of care into a redefinition of thinking as *therapeia*, belongs to the elaboration of a trans-formative sense. All this must also be seen against the backdrop of a melancholiform suffering under the unacceptable conditions in the absolute danger of the Entropocene. And let's not forget: Theodor Adorno, an author whose diagnosis he takes up early on, an early thinker of the coming barbarism that in Stiegler appears as the absence of epoch, an author whom he rereads time and again, explicitly highlights suffering as a critical affect. For Adorno, suffering precedes thinking. It is the product of a wrong society and unjust conditions. In *Negative Dialectics*, he writes: 'Where the thought transcends the bonds it tied in resistance – there is its freedom. Freedom follows the subject's urge to express itself. The need to lend a voice to suffering is a condition of all truth. For suffering is objectivity that weights upon the subject; its most subjective experience, its expression, is objectively conveyed.'[50] In a way, Stiegler's thinking has made suffering in the disruption speak, transformed it into a kind of conceptual stream for its healing interpretation – even if for him personally, the melancholiform aspect of thinking may have gained the upper hand over its therapeutic-healing-caring aspect. The first part of *The Age of Disruption* is entitled 'The *Epokhē* of My Life: Philosophizing So as Not to Go Mad'. And further on we read about 'the contemporary melancholy' that emblematically looms over the un-time of the disruption.[51] It points us to a surrender to the futility of trying to care for the world in the disruption, because in the barbarity of worldlessness and epochlessness, disruption always already forestalls any caring transformation, because the epochs of care have thus come to end and there is no more time of transformation left, no more time to be given. If anything, only a caring thinking that evades the melancholiform of the disruption is able to give another turn to what is without epoch and without world. For me, the great question is how this caring thinking must go and muster the strength to territorialize, against the terrible deterritorializations of the disruption but also against the reterritorializations of the ressentiments that only serve once again to amplify the general carelessness.

Translated by Nils F. Schott

Notes

1. Maurice Blanchot, *The Infinite Conversation*, trans. Susan Hanson, Minneapolis: University of Minnesota Press, 1993, 264.
2. The difference between the *who* and the *what* organizes the second part of Stiegler's *Technics and Time, 1: The Fault of Epimetheus*, trans. Richard Beardsworth and George Collins, Stanford: Stanford University Press, 1998, which gives a reading of Heidegger's *Being and Time*, trans. John Macquarrie and Edward Robinson, New York: Harper, 1962 and, in particularly, elaborates the 'the hypothesis of a *technological time* (the time of *what*), constitutive of the temporality of the *who*' (Stiegler, *Technics and Time*, 1, 210, Stiegler's emphases).
3. Daniel Ross makes a convincing case for this delineation in the introduction to his edition and translation of a collection of texts by Stiegler; see Daniel Ross, 'Introduction', in Stiegler, *The Neganthropocene*, 22–6.
4. Blanchot, *The Infinite Conversation*, 265–6, 268, 270, my emphasis.
5. See Stiegler, *Technics and Time, 1*, 221, 228.
6. See Edmund Husserl, *Ideas Pertaining to a Pure Phenomenology and to a Phenomenological Philosophy. First Book: General Introduction to a Pure Phenomenology*, trans. Fred Kersten, Volume 2 of *Collected Works*, The Hague: Nijhoff, 1983, §§31–2, 57–62.

7   Bernard Stiegler, *Technics and Time, 2: Disorientation*, trans. Stephen Barker, Stanford: Stanford University Press, 2009, 7.
8   Stiegler, 'Programs of the Improbable, Short Circuits of the Unheard-of' (1986), *Diacritics* 42/1 (2014): 70–109.
9   Stiegler, 'Programs of the Improbable, Short Circuits of the Unheard-of', 84–5.
10 Stiegler, 'Programs of the Improbable, Short Circuits of the Unheard-of', 86.
11 Stiegler makes the point in discussing the interruption of writing that opens up a new epoch – the time of history. This entire reflection on the technological condition is already taking place in light of the new technological condition of the informational or digital *epokhē*. According to Stiegler – and this shows the virulence of his enterprise and indicates the horizon of the care his thinking increasingly comes to express – this new condition, for all its deterritorializing and detemporalizing effects, proves to be not very productive when it comes to the doublings required for opening a new epoch, to be rather inhospitable towards the improbable, the unheard-of, the extraordinary that creates a new epochal program and threatens to be eliminated. See Stiegler, 'Programs of the Improbable, Short Circuits of the Unheard-of', 94.
12 Bernard Stiegler, 'How I Became a Philosopher', in Bernard Stiegler, *Acting Out*, trans. David Barison, Daniel Ross, and Patrick Crogan, Stanford: Stanford University Press, 2009, 1–35, 22, Stiegler's emphasis.
13 See Stiegler, *Technics and Time, 1*, 185–203.
14 See Stiegler, *Technics and Time, 1*, 231–7, esp. 235, where 'linear and phonological writing' is said to be 'a programmatic *epokhē*'.
15 As Stiegler himself saw: in 'What Is Called Caring?' he mentions that the concept of 'doubly epokhal redoubling' 'supports all the work carried out after *Technics and Time, 1*'; see Bernard Stiegler, 'What Is Called Caring?', in Bernard Stiegler, *The Neganthropocene*, ed. and trans. Daniel Ross, London: Open Humanities Press, 2018, 188–270, 302 fn. 396.
16 I would like to highlight here the absolutely originary difference Stiegler describes between problem and question: 'the problem is what provokes an exosomatic shock; the question is what seeks to care of it – where caring [*panser*] is called thinking [*penser*]' (see Bernard Stiegler, *Qu'appelle-t-on panser? 1: L'immense régression*, Paris: Éditions Les Liens Qui Libèrent, 2018, 71–2).
17 Stiegler, *Technics and Time, 2*, 60–1.
18 In the end, Stiegler – a monstrous suspicion – considers objective deconstruction to be caught up in the entropocenic movement of denoetization and total proletarization and to be entangled – at least where, for example in forgetting the problem of care, it refuses to think the techno-logical – in the ideology of disruption. See Stiegler, 'What Is Called Caring?', 197–8.
19 Bernard Stiegler, *The Age of Disruption: Technology and Madness in Computational Capitalism*, followed by *A Conversation about Christianity with Alain Jugnon, Jean-Luc Nancy and Bernard Stiegler*, trans. Daniel Ross, Cambridge: Polity, 2019, 14–15.
20 Stiegler, 'What Is Called Caring?', 191.
21 Stiegler, 'What Is Called Caring?', 191–2.
22 See Bernard Stiegler, 'The Anthropocene and Neganthropology', in Stiegler, *The Neganthropocene*, 24–50, 36.
23 Stiegler, 'What Is Called Caring?', 198. Derrida examines the 'privilege of the question' in Heidegger in Derrida, *Of Spirit: Heidegger and the Question*, trans. Geoffrey Bennington & Rachel Bowlby, Chicago: University of Chicago Press, 1989, esp. 15–16, where he also points to the 'unquestioned possibility of the question' and to 'the

privilege of the question having some relation already, always, with this irreducibility of technology'. In Stiegler's pharmacology of technics, every technical object is a *pharmakon*, that is, simultaneously a poison and a remedy: 'The *pharmakon* is at once what *enables* care to be taken and that *of which* care must be taken – in the sense that it is necessary *to pay attention*: its power is *curative to the immeasurable extent* [*dans la mesure et la démesure*] that it is also *destructive*' (see Bernard Stiegler, *What Makes Life Worth Living: On Pharmacology*, trans. Daniel Ross, Cambridge: Polity, 2013, 4). For the sketch of a 'pharmacology of the question', see Stiegler, *What Makes Life Worth Living*, 99–133.

24 See Stiegler, 'What Is Called Caring?', 214–15.
25 Stiegler, 'What Is Called Caring?', 215.
26 Stiegler, 'What Is Called Caring?', 209.
27 Stiegler, *The Age of Disruption*, 15.
28 The formula 'an "epoch" of the absence of epoch' (with the addition 'or the epoch of program industries') first appears in *Technics and Time, 1*, 213, but is elaborated only later.
29 The worldless at issue here concerns strictly speaking both meanings of the concept of world that Badiou aptly distinguished: as 'horizon of sense for all experience' and 'as logical figure, as consistent distribution of what appears'; see Alain Badiou, *Images du temps présent 2001–2004*, Paris: Fayard, 2014, 75 and 70. Stiegler's concept of world remains to be spelled out. On Granel's thinking of worldlessness that intertwines Husserl's problematization of modernity's release of a process of infinitization with Heidegger's critique of *Gestell* as deworlding technical world-making and Marx's description of Capital-Form, see my 'Die Problematik Granels', in Gérard Granel, *Die totale Produktion: Technik, Kapital und die Logik der Unendlichkeit*, ed. Erich Hörl, trans. Laura Strack, Vienna: Turia+Kant, 2020, 7–37.
30 Stiegler, *Qu'appelle-t-on panser? 1*, 51.
31 I have developed this Time-Form under the heading of the 'Time-Form of Environmentality' in 'Critique of Environmentality: On the World-Wide Axiomatics of Environmentalitarian Time', trans. Nils F. Schott, in ed. Erich Hörl, Nelly Pinkrah & Lotte Warnsholdt, *Critique and the Digital*, Zurich: Diaphanes, 2021, 109–44. In fact, the concept of 'tertiary protention' was originally fleshed out by Yuk Hui (in Yuk Hui, *On the Existence of Digital Objects*, Foreword by Bernard Stiegler, Minneapolis and London: University of Minnesota Press, 2016). Hui supplements Stiegler's pivotal conceptualization of tertiary retention to emphasize that under conditions of digitality, 'technology becomes a significant function of imagination', that 'imagination itself is no longer the imagination of the subject but rather shifts from subject to algorithms and digital objects' (221–2). This shift opens up another order of experience. In working on the computational *Gestell* Stiegler adopts this concept. Stiegler himself at least mentioned the concept of tertiary protention already in 2012 when speaking about money as 'element of grammatization and as tertiary protention, that which allows time (the time of the protentions in which belief essentially consists) to be trans-formed into an exchangeable and storable quantity', but he did not elaborate the concept. See Bernard Stiegler, *States of Shock: Stupidity and Knowledge in the 21st Century*, trans. Daniel Ross, Cambridge: Polity, 2015, 144. Stiegler himself concedes in his foreword to Hui's book: 'Yuk Hui tries to literally refound the question of time [...] by introducing his own fundamental concept of tertiary protention', which, as a concept 'echoes what I have tried to think as tertiary protention.' (Hui, *On the Existence of Digital Objects*, x)

32 Stiegler, 'What Is Called Caring?', 194, and Stiegler, *The Age of Disruption*, 15–17.
33 Stiegler, *Qu'appelle-t-on panser? 1*, 57.
34 Stiegler, *Qu'appelle-t-on panser? 1*, 58.
35 Stiegler, 'What Is Called Caring?', 212.
36 Stiegler, *Qu'appelle-t-on panser? 1*, 128.
37 Heidegger defines care temporally as 'ahead-of-itself-already-being-in (a world) as Being-alongside (entities encountered within-the-world)' in §41 and §65 of *Being and Time*.
38 Stiegler, 'What Is Called Caring?', 268. I cannot go into the details of the problem of care here, but the main point is this: Stiegler focuses his late readings of Heidegger on the problem of care and points out that Derrida's readings of Heidegger do not pay attention to it. This forgetting of care would also point us to the problematic place of deconstruction in the disruption: might it not, ultimately, 'merely gather […] up this *defabrication* of preceding noetic circuits by the *disruptive fabrications* of […] capitalism' (Stiegler, 'What Is Called Caring?', 198, Stiegler's emphases)?
39 Stiegler, *Qu'appelle-t-on panser? 1*, 77; see also 149–59.
40 Stiegler, 'The Anthropocene and Neganthropology', 36.
41 Stiegler, 'What Is Called Caring?', 200, 202.
42 Stiegler, 'The Anthropocene and Neganthropology', 45.
43 See Stiegler, 'What Is Called Caring?', 206–9. On the concept of a 'general organology', see for instance Stiegler, 'The Anthropocene and Neganthropology', 43–5.
44 In her essay 'The Care of the Possible' (*Cultural Politics* 12/4 (2016): 339–54) Verena Andermatt Conley notes a reinterpretation of the problem of care around the question of the possibility and of new ways of becoming, an immanence-philosophical reinterpretation that equally concerns humans and non-humans. Stiegler would not only have to be included in this operation of renewal, he might even be said to articulate its historical grounds and realize their epochality.
45 Martin Heidegger, *What Is Called Thinking?*, trans. and with an introduction J. Glenn Gray, New York: Harper and Row, 1968.
46 See Erich Hörl, 'Heidegger and Cybernetics', in Erich Hörl, *Sacred Channels: The Archaic Illusion of Communication*, trans. Nils F. Schott, Amsterdam: Amsterdam University Press, 2018, 299–317.
47 Jacques Derrida, *Of Grammatology*, trans. Gayatri Chakravorty Spivak, corrected ed., Baltimore: Johns Hopkins University Press, 1997, 8.
48 Jacques Derrida, *Margins of Philosophy*, trans. Alan Bass, Chicago: University of Chicago Press, 1982, 7 (translation modified), 22.
49 Stiegler, *Qu'appelle-t-on panser? 1*, 163–4.
50 Theodor W. Adorno, *Negative Dialectics*, trans. E. B. Ashton, London: Routledge and Kegan Paul, 1973, 17–18.
51 Stiegler, *The Age of Disruption*, 21. And let's not forget: the 'contemporary melancholy' that appears here, at the end of an oeuvre, ultimately responds to the 'the eternal melancholy of the *genos anthropos*' Stiegler had already noted in *Technics and Time, 1*, 190.

Chapter 10

DIFFÉRANCE AND EPOCHALITY: STIEGLER'S *TOURS*

Donovan Stewart

Stiegler's challenge

Bernard Stiegler's texts challenge. Marked by pace, aggression and an abundance of concepts, they pair a challenging style with a deeper challenge for *panser*[1] in the Anthropocene epoch, understood as the result of an overproduction of entropy.[2] In response, Stiegler offers a twofold challenge: identify and re-form entropic structures, and create futures that delay and differ from this universal tendency. Understood more broadly, Stiegler's texts are a challenge *of* justice – its force shapes the contours and turns of his thought.[3]

How does Stiegler respond to these challenges that his texts bring forth with such urgency? Are his diagnoses of and responses to the present age successful? And how is success to be determined against the infinite challenge of justice? At first glance, his descriptions are provocative. 'Our' times are 'the age of disruption', the 'final stage of the Anthropocene', 'accomplished nihilism', when 'we all [...] find ourselves *thrown into* and thrown *out* by the epoch of the absence of epoch'.[4] This style is significant, especially because of its proximity to the thought of Martin Heidegger, who also diagnoses the present age as being threatened by a certain technical epoch, a proximity that Stiegler acknowledges in 2016:

> We cannot but be struck by the following sentence [... from Heidegger's *Discourse on Thinking*]: 'The approaching tide of technological revolution in the atomic age could so captivate, bewitch, dazzle, and beguile man that calculative thinking may someday come to be accepted and practiced *as the only* way of thinking'. [Stiegler continues:] It is clear that this is the point at which we have now arrived. 'Then there might go hand in hand with the greatest ingenuity in calculative planning and inventing indifference toward meditative thinking, total thoughtlessness.'[5]

Although Stiegler's project begins precisely as a turn away from a 'Heideggerian' technophobic heritage, as the passage above suggests, the difference between the two thinkers can occasionally become difficult to locate, especially when Stiegler turns his attention towards the present epoch.

Stiegler's descriptive style also attests to an ethical urgency that animates his texts and political projects – an urgency that may help explain the occasionally fast nature of his texts, as well as the potentially hasty, *potentially pragmatic*, determination of the present age as being exceptional, of a single we, world, and (absence of) epoch, that shares one, probable, total end.[6] This hyperbolic tendency draws accusations of sensationalism, and perhaps legitimately so, yet philosophy and theology are no longer the only discourses concerned with a total end, having been joined by the sciences which increasingly warn of the irreversible threat to life that is posed by the Anthropocene. Taking this ecological crisis seriously, how to best appropriate Stiegler's concerns of the total end while maintaining his challenge of justice to cultivate differences, new horizons, and the deferral of entropy? Does his writing of the present epoch effectively respond to this challenge? Or does it obstruct thought by drawing it towards the total end?

Although Stiegler's work is not exhausted by this apocalyptic concern, it does gravitate towards it; moreover, I suggest that this tendency hinders his political theory and does not hold up against his thinking of technical *différance*.[7] This tendency grows over time, but it is already present in his first major work, *Technics and Time, 1: The Fault of Epimetheus*, which offers a novel understanding of technics as *différance* as well as a theory of the exceptional danger of the present epoch.[8] A tension between these two turns of thought, *différance* and epochality, draws Stiegler towards a metaphysical understanding of technics as a totalizing force of calculation. However, his theory of technical *différance* destabilizes the rigour of any determinate (absence of) epoch, upon which this understanding relies, and leads beyond the thought of closure, towards the need to cultivate technical idiomaticity, while remaining vigilant about the dangers posed by totalizing forces today.

Two turns: Différance *and epochality*

Stiegler's two turns of thought can be summarized by two terms: technical *différance* and epochality. These two turns and two terms are intimately related, stemming from the same texts and meditations, yet are potentially irreconcilable, opening onto significantly different paths for thought. Stiegler first turns from traditional pejorative determinations of technics by developing an account of the originary technicity of human being, according to which technics is the *différance* of human being that gives time. Stiegler's second turn pivots from this general theory, to develop an epochal history of technical life, focusing on the present epoch in which technical development is no longer guided by discursive politics, but rather by an automated market, introducing the threat that technics could foreclose *différance* and prevent the coming of an incalculable future. However, this concern may reintroduce a traditional understanding of technics that the first turn overcame. These two turns occur side by side in *Technics and Time, 1*, which offers a framework for the following two volumes of the series and Stiegler's *oeuvre* as a whole, and they arise as a turn from, and a return to, Heidegger's determination of modern technics as *Gestell*.

1st turn: The calculable opening of the incalculable

Stiegler deconstructs the traditional determination of technics,[9] as a perversion to 'what is proper to humanity' by demonstrating that human properties are possible through technical supplementation alone.[10] Human being and technics move in a 'transductive' dynamic – a co-constitution of organic and inorganic, *who* and *what*, in which 'the *what* invents the *who* just as much as it is invented by it'.[11] Thus, human life is not purely biological, spiritual, living or even human, but firstly a site of technical exchange, contaminated by inanimacy, exteriority, alterity and death. For Stiegler, this 'originary technicity' strips the human of any transcendental dimension that would be unscathed by technics, its most intimate properties being, at least in a certain sense, determined by technical exteriority. Technics are the external organs of human being, constitutive of cognitive, emotional and physical health, as well as memory.

Technical life inscribes memories in the world, and thus, 'far from being lost with the living when it dies, [technical life] conserves and sediments itself, pass[ing] itself down in "the order of survival" [*survivance*] and to posterity as a gift as well as a debt, that is, as a destiny'.[12] This technical form of memory permits life to live beyond its (traditionally determined temporality of) pure immediacy, and Stiegler emphasizes that such externalized memory is essential for technical life, which becomes itself by appropriating the inscribed memories of others as a heritage. Importantly, this process of inheritance is not an identical reproduction of the past, but an exappropriation that introduces loss and differentiation. For Stiegler, this composition of life and technics is a *différance*, an interplay that defers a past, creates a calculable difference and opens onto an incalculable future. He argues that this technical *différance* has been repressed by Western thought as can be seen in Heidegger's work on time, which Stiegler identifies as an instance of metaphysics' degradation of technics, as well as an early articulation of the originarily technical dimension of temporality.

In *The Concept of Time*[13] and *Being and Time*,[14] Heidegger argues that Western metaphysics has understood time on the basis of 'datability' or temporal presence. For Heidegger, this misconstrues its essence which, in a more originary sense, is an auto-affective operation of the future that gives presence to be thought:

> [It is only] because temporality is ecstatico-horizonally constitutive for the clearness of the 'there', [that] temporality [is] always primordially interpretable in the 'there' [... as] that which has been interpreted and is addressed in the 'now' – [which] is what we call 'time'.[15]

Heidegger understands authentic temporality as a pre-technological movement that is subsequently ordered according to the techno-logic of the clock. For Stiegler, Heidegger's gesture follows the traditional prioritization of pre-technological purity by creating an 'opposition between the time of technical measurement [...] and authentic time', that thus presupposes technics to be a malevolent force of calculation.[16]

However, Heidegger's understanding of historicity simultaneously indicates how temporality is in fact made possible through technical supplementation alone. Stiegler suggests that if *Dasein* 'is its past', which 'discloses' and 'regulates' the 'possibilities of its Being', and if this past is necessarily technical, then technics must shape *Dasein*'s 'authentic' future.[17] Stiegler reads Heidegger's passage: *Dasein*'s 'uttermost and ownmost possibility is coming back understandingly to [its] ownmost "been"', as an early formulation of how *Dasein*'s futurity arises from an earlier relay through history, meaning that there is no incalculable future prior to technical calculability, public inauthenticity, faceless archives and the time of the clock.[18] For Stiegler, this relay is a *différance* that creates differences from the already-there, defers from (temporizes) the present, and opens onto an indeterminate future:

> The gift of *différance* is technological because the individual constitutes itself from out of the possibilities of the One, from the relation with one another each time allowed for by the particular technological set up. [...] The calculation of time is thus not a falling away from primordial time, because calculation, *qua* the letter-number, also *actually* gives access in the history of being to any *différance*.[19]

In this way, Stiegler turns beyond metaphysical determinations of technics, technophobic and technophilic alike. Neither individual technics nor technics as such can be exhaustively determined by an essence, but must be understood as *pharmaka* – toxic and/or curative forces in need of *panser* – a simple insight that transforms the philosophy and politics of technics. Lacking recourse to the rejection of technics, and refusing to acquiesce to technical developments as such, Stiegler demands for an attentiveness to contemporary technics and the worlds they form, and an awareness of how technical life comports itself towards 'future protentions' or desired futures that shape the present.[20] *Panser* is at once the evaluation of future protentions, and the gathering of technics into line with their law, forming memories for and of the future. Stiegler's understanding of technical *différance* allows for this conception of futurity which overcomes a metaphysical tradition that has determined technics as a force of closure.

That the thought of technical *différance* arises through an encounter with Heidegger is no coincidence, for the ramifications of his understanding of technics necessitate a radical turn. For Heidegger, what may seem to be an innocent misunderstanding of time in 'datability' is no accident at all, but an instance of a broader becoming-technical that characterizes the present epoch. Heidegger places 'datable' time in a larger constellation with the inauthenticity of the city, the degradation of *logos* to *ratio*, and the obfuscation of the *Seinsfrage* by discourses concerning beings; all of which are effects of technology's essence – *Gestell*.[21] Technics understood as *Gestell*, 'transposes [*heraussetzt*] all that is out of its previous essence',[22] ordering beings as *Bestand* or 'standing-reserve': 'Everywhere everything is ordered to stand by, to be immediately at hand, indeed to stand there just so that it may be on call for a further ordering'.[23] For Heidegger, *Gestell* strips beings of their *Geschlecht*: their sexual, gendered, ethnic, generational and national difference, erasing their distinct possibilities of 'hav[ing] a kind'.[24] 'The piece of standing reserve [*Bestand*] does not even share itself with its own kind in the standing reserve. The standing reserve is much more that which has been shattered [*Zerstückte*] into the orderable'.[25] *Gestell* neutralizes *Geschlechter*; and for Heidegger this is the movement of Western nihilism, which culminates in the total end in which beings, including human being, are ordered as interchangeable pieces in an undifferentiated mass.[26] Although Stiegler avoids this apocalyptic logic by thinking technics as *différance*, his second turn presents an epochal history of technics that may re-summon this logic of *Gestell*, along with the fear of an impending 'blank' *Geschlecht*.

2nd turn: Epochality and 'Our Geschlecht'

Stiegler's originary technicity thesis presents the differential play of human being and technics in a way that allows technical *différance* to be thought independently of specific technical systems. This seems to be the case when Stiegler writes:

> *Différance* is neither the *who* nor the *what*, but their co-possibility, the movement of their mutual coming-to-be, of their coming into convention. [...] *Différance* is below and beyond the *who* and the *what*; it poses them together, a composition engendering the illusion of an opposition.[27]

Even though *différance* takes place in relation to technics, it cannot be brought under control in order to preserve a given order of things. As Derrida writes, such attempts at mastery are automatically exposed to their own deconstruction:

> Deconstruction takes place, it is an event that does not await the deliberation, consciousness, or organisation of a subject, or even of modernity. It deconstructs itself. It can be deconstructed. [*Ça se déconstruit.*] The 'it' [*ça*] is not here an impersonal thing that is opposed to some egological subjectivity. It is in deconstruction (the *Littré* says, 'to deconstruct itself [*se déconstruire*] [...] to lose its construction'). And the '*se*' of '*se déconstruire*,' which is not the reflexivity of an ego or of a consciousness, bears the whole enigma.[28]

Différance is always underway; insofar as something *is*, it is already infected by its operation. 'It is in deconstruction'.

Stiegler's understanding of the 'enigma' of the '*se*' marks what I call his second turn. Diverging from Derrida, he seeks to *determine* this source of heterogeneity. Whereas Derrida's concepts of 'trace', 'alterity' and 'Other', are temporary, 'crossed-out' place holders that gesture towards this operation without being its truth or guarantor, for Stiegler, there is a delimitable source of the possibility of *différance* – technics – framed by *epochs*.

> Nothing can be said of temporisation [*différance*] that does not relate to the [...] structure put in place each time, and each time in an original way, by [...] the memory supports that organise successive epochs of humanity: that is, technics – the supplement is elementary, or rather elementary supplementarity *is* (the relation to) time (*différance*).[29]

For Stiegler, *différance* is made possible by epochs that are formed through successful negotiations with technical milieux, between a *who* and a *what*. Epochs are cultural 'programs' that give people a 'heritage' or *Geschlecht* with a 'non-genetic memory' and 'collective future'.[30] As shown above, this gift of heritage provides technical life with its 'possibilities of differentiation and individuation', but more fundamentally, here Stiegler suggests that the technical specificity of an epoch provides the very possibility of *différance*, and implicitly, can take it away.[31]

Stiegler thus argues that Derrida's formulations remain too transcendent because they do not base the possibility of *différance* on technics, and fail to account for the 'regimes of *différance*, [... its] kingdoms, ages, eras (both geological and theological) and epochs [... that must] be drawn out'.[32] This marks Stiegler's dramatic second turn which diverges from Derrida, for whom a 'kingdom of *différance*' makes no sense, since *différance* exposes any such closure to its outside. As Derrida writes in 'Différance':

> Not only is there no kingdom of *différance*, but *différance* instigates the subversion of every kingdom. Which makes it obviously threatening and infallibly dreaded by everything within us that desires a kingdom, the past or future presence of a kingdom. [...] Perhaps [...] the history of Being [...] is but an epoch of the *diapherein*. Henceforth one could no longer even call this an 'epoch', the concept of epochality belonging to what is within history as the history of Being.[33]

In light of this passage, in a certain way Stiegler attempts to draw *différance* back into Being and expose it to its 'most extreme possibility' – its death – or complete effacement. Indeed, his understanding of epochality is motivated by the fear, and perhaps fascination, of this extreme possibility, which he argues is already the reality today.

If Stiegler turned from Heidegger's path concerning temporality, epochality draws them back together. For Stiegler, speculative capitalism develops technics too quickly for epochs to form ('redouble'), leading to a state of 'contemporary disorientation': 'the experience of an incapacity to achieve epochal redoubling' caused by 'current prosthetics [that] act as an obstacle to [redoubling]'.[34] Thus

Stiegler insists that at present there is no epoch, meaning that the *who* and the *what* cannot compose, heritage is not shared, *différance* is 'concealed in an essential manner',[35] and our *Geschlecht* is left with no future:

> The generation of today's 'time', our *Geschlecht*, says flatly: *no future*. What is affirmed here at the same time as it is refused? Does this slogan mean that there is no *différance* or no longer any *différance* in the extrapolation of the present as *Gegenwart* – that in 'real time', which is nothing but this extrapolation, *there can be no future?*
>
> If such a question invited a response, indeed an affirmative one, it would not mean simply saying that *tekhnē* conveys the power that propels the *who* to its falling [...] precisely because it is *tekhnē* that *gives différance*, that *gives time*.[36]

Stiegler tries to maintain distance from Heidegger by stressing that the problem today is not technics as such, but the present technical (lack of) epoch, yet their determinations of the present become nearly indistinguishable: modern technics prevents the coming of an incalculable future and strips beings of their differences, forming a 'blank *Geschlecht*'. '"Our present" generation no longer believes that it belongs to anything; in its eyes, no longer is anything wanting: the default has become general [*ça fait défaut*]. The Blank "*Geschlecht*" is the generation of default.'[37]

Stiegler maintains this determination of the present throughout his writings: a lack of epoch prevents the work of *différance*, leaving us guided solely by the 'negative teleology' of mass culture.[38] For Stiegler, our grim reality amounts to the 'age of the concretization of what Heidegger called *Gestell*';[39] in which it is as if there were no future, and as if modern technics had sublimated *différance* – as if.

The challenge of justice: Sovereign responses, hyperethical responses

Stiegler's two turns present two incompatible thoughts: on the one hand, human being and technics compose in a technical *différance* that gives calculable differences and the incalculability of the future; and, on the other, *différance* is today foreclosed by a toxic technical milieu. This thesis depends on the claims that *différance* is grounded on technics and that contemporary technics prevent epochs from forming. However, I suggest that this tension most fundamentally is the result of a decision, Stiegler's sovereign decision to declare a present absence of epoch.[40]

1. Stiegler's determination assumes that there could be a present epoch to be claimed as 'ours'. How rigorous is this unity? After all, Stiegler demonstrates that every epoch and every *Geschlecht*, along with their shared beliefs, futures and pasts, are 'inevitable phantasms'.[41] Is not the delimitation of an epoch also phantasmic, also based on an 'as if'?[42] How else can one determine the borders of a phantasmic epoch – can (an absence of) a phantasm be delimited *as such*?

To declare epochal absence is to claim to account for every single other, and to know that phantasms are not present without remainder. But has there ever been, for certain, no (unperceived) epoch, a verifiable lack of heritage, or a confirmed void of collective futures?
2. Of course not – but of course, there is always the possibility and the necessity for such declarations, erected by sovereign forces that organize the future *as if* it were safe from the operation of *différance* which is always already underway.
3. Is there not 'today' a multiplicity of epochs that put the unity of a single 'today' into question?[43] Does not Stiegler follow Heidegger in dismissing contemporary epochs as non-epochs that introduce discord to past, purer epochal gatherings, when in fact they are the *différance* of past *Geschlechter*?

Simply put, there never has been, nor ever will be, a true epoch, meaning that every search for a lack of epoch will necessarily result in success. This latter search characterizes Stiegler's tendency towards the total end that turns him away from contemporary techno-diversity, and towards the phantasm of a coming blank *Geschlecht*. However, in light of these *tours*, is Stiegler a sovereign thinker of his unique age and its end, who ignores those differences which exceed his measures? Perhaps.

And yet, Stiegler's work is fundamentally *of* justice, shaped as a prolonged response to its paradoxical call. As Derrida describes, in order to respond to justice's pure demand, one must commit, and always already commits, an inexplicable, unforgivable injustice in deciding to act *as if* it were the right time and place, when in fact it is always too late and too early. In order to be ethical, one must make a decision against pure ethics, a decision that is necessarily anterior to reason and moral calculation. There is no explanation possible. For Derrida, such a 'hyperethical'[44] decision is the aporetic, unjust origin of any just response to the challenge of justice. A necessary, sovereign violence. Of justice.

I can respond to the one (or to the One), that is to say to the other, only by sacrificing to that one the other. I am responsible to any one (that is to say to the other) only by failing in my responsibilities to all the others, to the ethical or political generality. And I can never justify this sacrifice, I must always hold my peace about it. Whether I want to or not, I will never be able to justify the fact that I prefer or sacrifice any one (any other) to the other. I will always be in secret, held to secrecy in respect of this, for nothing can be said about it. What binds me to singularities, to this one or that one, male or female, rather than that one or this one, remains finally unjustifiable (this is Abraham's hyperethical sacrifice), as unjustifiable as the infinite sacrifice I make at each moment.[45]

Is not Stiegler's second turn, his decision regarding *différance*, *of* this challenge?[46] Are not his hyperbolic determinations of the present the result of a sovereign-hyperethical decision, a necessary failure of justice for justice?

How dare you? There is no response possible. *How dare you?* This is Greta Thunberg's *parrhesia* around which Stiegler's final text revolves, a text that interrupts the series, *What Is Called Caring?* with this refrain: *How dare you?*[47] This challenge gives voice to the force of justice that demands for, and overturns, every ethical decision – *There is no time, I must respond, I must sacrifice, I have failed. Again!* These *tours* shape Stiegler's texts, which understand that justice requires sacrifice and the courage for sovereign-hyperethical violence. How shall we inherit *this* challenge?

I conclude by way of a memory, Stiegler's, of his doctoral defense recalled twenty years later, a memory that brings forth the fundamental incommensurability of every response to justice:

> Derrida concluded his introductory remarks on my thesis, which he supervised, in front of a jury chaired by Jean-Luc Marion, by asking me the question, 'What is it you're afraid of?' In the course of the weeks and months that followed, I could not help but think that, in this way, Derrida was both downplaying the gravity of the situation I was attempting to describe as the uncontrolled extension of retentional technology, and making a show of being someone 'who is not afraid'. The question I was raising, however, had nothing to do with fear [*peur*] in a way connected to cowardice, but to do with worry [*crainte*], in a way that requires courage.[48]

Notes

1. *Panser* [to dress/care for a wound] plays on *penser* [to think], indicating the therapeutic dimensions of thought: 'We ought to conceive thinking [*pensée*] as that which consists in caring [*panser*], that is, in taking care' (Bernard Stiegler, *The Nanjing Lectures 2016–2019*, trans. Daniel Ross, London: Open Humanities Press, 2020, 78). Cf. Daniel Ross' n. 171 in Bernard Stiegler, *The Neganthropocene*, trans. Daniel Ross, London: Open Humanities Press, 2018, 284–5.
2. Stiegler, *The Nanjing Lectures*, 10: 'As a vast, systemic and extremely rapid process of increasing entropy, the Anthropocene necessarily leads to the destruction of all kinds of life, and firstly, of human life'.
3. This *of* is double genitive, meaning that 1) Stiegler's texts challenge us to do justice and, 2) they respond to a challenge from justice. This draws from Jacques Derrida's understanding of a 'hollow', unattainable, justice that exceeds law, yet demands for it. Stiegler writes: 'Law is even what affirms, beyond the law, and as its very promise, a justice that will never eventuate, which will therefore *never be cured* of injustice' (Bernard Stiegler, 'What Is Called Caring? Thinking beyond the Anthropocene', in Stiegler, *The Neganthropocene*, 188–270, 200).
4. Bernard Stiegler, *The Age of Disruption: Technology and Madness in Computational Capitalism,* followed by *A Conversation about Christianity with Alain Jugnon, Jean-Luc Nancy and Bernard Stiegler*, trans. Daniel Ross, Cambridge: Polity Press, 2019, 7; Bernard Stiegler, 'Elements of Neganthropology', in Stiegler, *The Neganthropocene*, 76–91, 81; Bernard Stiegler, 'Escaping the Anthropocene', in Stiegler, *The Neganthropocene*, 51–63, 53; Stiegler, *The Age of Disruption*, 12.
5. Stiegler, *The Age of Disruption*, 282; Martin Heidegger, *Discourse on Thinking*, trans. John M. Anderson & E. Hans Freund, New York: Harper and Row, 1969, 56.

6   On the question of speed in Stiegler's style, cf. Richard Beardsworth, 'Technology and Politics: A Response to Bernard Stiegler', in *Stiegler and Technics*, ed. Christina Howells & Gerald Moore, Edinburgh: Edinburgh University Press, 2013, 208–24, 223.
7   'Technical *différance*' refers to the differential operation that composes human being and technics that is discussed at length below.
8   Bernard Stiegler, *Technics and Time, 1: The Fault of Epimetheus*, trans. George Collins & Richard Beardsworth, Stanford: Stanford University Press, 1998.
9   'Technics' should be understood with the French '*technique*', which, Susanna Lindberg reminds, 'denotes a vast variety of phenomena that the English language divides into *technologies* and *techniques*' (Susanna Lindberg, 'Derrida's Quasi-Technique', *Research in Phenomenology* 46 (2016): 369–89, 370).
10 Stiegler, *Technics and Time, 1*, 13. Leroi-Gourhan demonstrates that the tool paradoxically constitutes human 'anticipatory capacities', allowing access to the future and the as-such of the *logos*. Cf. Stiegler, *Technics and Time, 1*, 141: 'Leroi-Gourhan in fact says that it is the tool, that is, *tekhnē*, that invents the human, not the human who invents the technical. Or again: the human invents himself in the technical by inventing the tool – by becoming exteriorised techno-logically'; see also Stiegler, *Technics and Time, 1*, 148–58.
11 Stiegler, *Technics and Time, 1*, 177.
12 Stiegler, *Technics and Time, 1*, 140.
13 Martin Heidegger, *The Concept of Time*, trans. Ingo Farin, New York: Continuum, 2011.
14 Martin Heidegger, *Being and Time*, trans. John Macquarrie & Edward Robinson, New York: Harper and Row Publishers, 2008.
15 Heidegger, *Being and Time*, 460. Emphasis modified.
16 Stiegler, *Technics and Time, 1*, 141. For a richer account of Heidegger's conception of facticity that supplements Stiegler's reading, cf. Tracy Colony, 'A Matter of Time: Stiegler on Heidegger and Being Technological', *The Journal of the British Society for Phenomenology* 41 (2010): 117–31.
17 Heidegger, *Being and Time*, 41.
18 Heidegger, *Being and Time*, 373.
19 Stiegler, *Technics and Time, 1*, 237.
20 This is of massive importance for Stiegler, increasingly formulated as the 'noetic dream' that can propose phantasmic, affective 'memories' that shape the present towards a desired future. Cf. Stiegler, '§86. The Political Function of Dreaming' (in Stiegler, *The Age of Disruption*, 193–6) and Stiegler, 'Conclusion: Let's Make a Dream' (in Stiegler, *The Age of Disruption*, 286–312).
21 Although Heidegger presents *Gestell* as a mode of *Seyn*'s '*Einblitz*' or flashing into presence, he nevertheless condemns it as a deficient form of revealing that leaves beings 'unsafeguarded' and 'truthless', cf. Martin Heidegger, 'The Turning', in Martin Heidegger, *The Question Concerning Technology and Other Essays*, trans. William Lovitt, New York: Harper and Row Publishers, 1977, 36–49, 46–7.
22 Martin Heidegger, *Bremen and Freiburg Lectures: Insight Into That Which Is and Basic Principles of Thinking*, trans. Andrew J. Mitchell, Bloomington: Indiana University Press, 2012, 4.
23 Heidegger, 'The Question Concerning Technology', in Heidegger, *The Question Concerning Technology and Other Essays*, 3–35, 17.
24 Heidegger, *Bremen and Freiburg Lectures*, 34. At once meaning 'sex', 'generation', 'stock' and 'race', *Geschlecht*, as Derrida cautiously demonstrates, is most essentially

'the gathering unity of [...] this multiplicity of significations, all the blows of which come in one single mark, one single word, a word that also says gathering (*Ge-*), in order to seal their consonance' (Jacques Derrida, *Geschlecht III: Sex, Race, Nation, Humanity*, trans. Katie Chenoweth & Rodrigo Therezo, Chicago: University of Chicago Press, 2020, 127). On '*Geschlecht*' as developed by Heidegger in *Unterwegs zur Sprache* (Pfullingen: Neske, 1959) and other texts, cf. Derrida's four part *Geschlecht* series, *Of Spirit: Heidegger and the Question*, trans. Geoffrey Bennington & Rachel Bowlby, Chicago: Chicago University Press, 1991 and David Farrell Krell's *Phantoms of the Other: Four Generations of Derrida's Geschlecht* (Albany: State University of New York Press, 2015).
25 Heidegger, *Bremen and Freiburg Lectures*, 34.
26 For the wider, dangerous implications of 'Heidegger's obsessed and obsessional logic of annihilation' and apocalypse, cf. Marcia Sá Cavalcante Schubach, 'Apocalypse and the History of Being', in Mårten Björk & Jayne Svenungsson (ed.), *Heidegger's Black Notebooks and the Future of Theology*, Cham: Palgrave Macmillan, 2017, 191–210, esp. 202–3.
27 Stiegler, *Technics and Time, 1*, 141.
28 Jacques Derrida, 'Letter to a Japanese Friend', trans. David Wood & Andrew Benjamin, in David Wood & Robert Bernasconi (ed.), *Derrida and Différance*, Coventry: Parousia Press, 1985, 1–5, 3.
29 Stiegler, *Technics and Time, 1*, 183.
30 Stiegler, *Technics and Time, 1*, 55.
31 Stiegler, *Technics and Time, 1*, 236.
32 Stiegler, *The Age of Disruption*, 281.
33 Jacques Derrida, 'Différance', in Jacques Derrida, *Margins of Philosophy*, trans. Alan Bass, Chicago: Chicago University Press, 1982, 3–27, 22. Emphasis modified.
34 Bernard Stiegler, *Technics and Time, 2: Disorientation*, trans. Stephen Barker, Stanford: Stanford University Press, 2009, 7.
35 Stiegler, *Technics and Time, 1*, 225.
36 Stiegler, *Technics and Time, 1*, 220.
37 Stiegler, *Technics and Time, 1*, 228.
38 Stiegler, *The Age of Disruption*, 12.
39 Stiegler, *The Nanjing Lectures*, 10.
40 Following Derrida, the 'sovereign gesture' is the moment of 'the proper appropriating itself, the proper *positing* itself': a violent, necessary forgetting of differences that allows for a decision (Jacques Derrida, *The Beast & the Sovereign, Volume 1*, ed. Michel Lisse, Marie-Louise Mallet, and Ginette Michaud; trans. Geoffrey Bennington, Chicago: University of Chicago Press, 2009, 192).
41 Bernard Stiegler, *Technics and Time, 3: Cinematic Time and the Question of Malaise*, trans. Stephen Barker, Stanford: Stanford University Press, 2011, 106.
42 On the phantasm and epochality, cf. Reiner Schürmann, *Broken Hegemonies*, trans. Reginald Lilly, Bloomington: Indiana University Press, 2003.
43 Which are not recuperations of past *Geschlechter*; the desire for mythological origins for 'ontological self-determination' repeats the xenophobic violence of the search for a phantasmic, pure community.
44 Jacques Derrida, *The Gift of Death*, trans. David Wills, Chicago: University of Chicago Press, 1995, 71.
45 Derrida, *The Gift of Death*, 71.

46 Another sovereign-hyperethical moment: without the ground of reason or morality, Stiegler prioritizes 'noetic' life over other forms of life. There is no explanation possible – a hyper-ethical decision is always wrong, always necessary. On Stiegler's 'decision' and nonhuman *différance*, cf. Tracy Colony, 'Epimetheus Bound: Stiegler on Derrida, Life, and the Technological Condition', *Research in Phenomenology* 4 (2011): 72–89.
47 Bernard Stiegler, *Qu'appelle-t-on panser? 2: La leçon de Greta Thunberg*, Paris: Les Liens Qui Libèrent, 2020.
48 Stiegler, *The Neganthropocene*, n. 555, 319.

Part IV

CREATIVE ORGANOLOGIES: WORKS OF INVENTION

Chapter 11

PHILOSOPHY THROUGH ACTING

Bart Buseyne

Bernard Stiegler was clearly a philosopher whose *passage to the act* spilled over into many a field: he co-founded several associations (including Ars Industrialis) and established an academy (Pharmakon.fr); he was invited to hearings by the Belgian senate, and was a member of the French Digital Council; he helped to develop a localized platform (Dassault) and to specify a new concept for social networks; he created software tools for reading texts (PLAO) and for viewing film (Timelines), and designed data architecture for the worldwide web; he assisted in setting up a contributory clinic (Saint-Denis) and in recasting town areas (Real Smart Cities); he participated in arts festivals (*Les Inattendues*), acted as a curator *(Mémoires du futur)* and worked with film directors (Ken McMullen), theatre producers (Valérie Cordy) and musicians (eRikm); he attended militant gatherings (6th Subversive Festival) and launched a call for a global economic peace treaty (Geneva 2020) …

The 'succession of roles'[1] Stiegler assumed on the terrains where the action takes place makes us wonder whether there might be some *intimate,* if not *secret* connection between his philosophical work and his involvement in techno-cultural experimentation and relevant policymaking: does his implication in these latter activities answer to a philosophical necessity of sorts? Our answer to that question is a clear yes. As the conjoined individuation of psychic and collective individuals, the life of the mind is framed by technical materializations that expose it from the inside. Since the mind's presence makes for a structure of 'absences', it is intrinsically 'fragile'.[2] This fragility not only calls for the elaboration of knowledge of the way in which spirit is entangled in a history of technical supports. It also encourages the development of practices of care that may facilitate the adoption of technical alterity in terms of the intensification of the co-individuation of 'the *Is* and the *we*'.[3] While there is no inside 'outside of the outside',[4] the technical movement of anamnesic life demands effort and attention. Only in this way may it be pursued, albeit always somewhat otherwise than anticipated. Owing to their essential exteriority, the practices that 'give place'[5] to anamnesic ways of life can only promise these an iteration and renewal that go beyond intent and recognition.[6] To care for styles of *noēsis* with the 'vocation' of one's heart is to experience how precious signifying practices are in the very *secret* of their intimacy, and to make these into 'the object of a veritable culture' that is a culture of their 'being *by default*'.[7]

Anamnēsis *as frequentation*

We take our cue from the lecture Stiegler presented in 2003 in response to Marianne Alphant's invitation to reflect on how he became a philosopher 'in the intimacy and secret'[8] of his life. In that talk, he only touched on the relation between philosophy and action in the field of technical application and relevant policy design. The narrative clearly indicates, however, the direction in which an answer may be found to the question of how noetic life calls from within to attend to 'its' outside. It is because the passage that takes him from a 'potential devotion' to philosophy 'into the act', is 'given place to by a milieu',[9] that philosophy, for him, comes to *flow over*[10] into an involvement with issues concerning the milieu of noetic co-individuation.

As Stiegler found out in prison, that milieu is made of hypomnesic supports. This discovery ties *anamnēsis* to a technical memory that gives rise to it. As noetic life is reliant on the *hypomnēsis* of 'the books read and words written',[11] its interior is nothing without the outside. Beginning to read and write, 'secreting' around him 'an *intimate* hypomnesic milieu at once *secret* and yet already *publishable*', Stiegler constituted 'a *world* that would become his philosophy'.[12] In thus reconstituting 'daily the artificial locality' of his 'writing and reading', he was able to 'continue to have a place'.[13]

In connecting thinking to the exteriority of 'its' milieu, Stiegler's anamnesic practice invites a description in terms of a passage to the limits of phenomenology.[14] As the milieu that is made by technical traces marks the 'retentional finitude'[15] of thought, it forms the limit of thinking: no 'author' precedes the instrumental supplementarity through which they may claim authority.[16]

That there is no inside outside of an outside 'originally assisting finitude',[17] entails that – as Stiegler realized – 'virtue is outside of [him]'.[18] Since reminiscence must be produced, *becoming* a philosopher requires endurance, if it is not to be made 'the object of the whole of one's care'.[19] Outside of 'a signifying material and a signifying practice', there simply is no significance, which hence demands 'the frequenting of a practice'.[20]

The condition of Anamnēsis

Taking up the properly philosophical question of reminiscence, Stiegler contests the way Plato structures the relation between *anamnēsis* and *hypomnēsis* as an opposition.[21] Casting the connection between the living spirit and the dead memory of technically organized matter in compositional terms, he reinvents philosophy as a practice of technically supported, therefore social rather than cerebral *passages to the act*.

At the origin of philosophy, with Plato, the question of origin opens itself as the question of reminiscence. For Stiegler, this question emerged as 'the question of the signifying practices that it was a matter of reinventing from out of the insignificance into which [he] had fallen'.[22]

That reminiscence is a question of practices means above all that it must be conceived by the one who articulates it. *Anamnēsis* cannot be received from the outside, it can only be produced by the subject inhabiting it. On account of its finitude, however, it can only be conceived with the help of an essentially exterior memory. The very default of living memory calls for mnemotechnical supports that first allow it to identify its objects and train of thought.

Plato has no difficulty recognizing how *anamnēsis* is supported by a predating memory that forms an 'already-there'[23] in which 'one finds both questions and answers'.[24] Yet, he does not consider this memory to be principally exterior as it consists of ideas beyond time and place. Stiegler, for his part, argues that the already-there must be apprehended in terms of technically organized matter.[25] It is the organization of matter into the support of *anamnēsis* that provides the condition of *anamnēsis*. *Anamnēsis* can consequently not be divorced from the artefacts through which we are able to think in the first place. After all, the very thought processes that lie behind Plato's invention of the ideas are made possible by inscriptions. If technics is the source of living memory, then the interior cannot be looked at as separate from its exteriorization into technically organized matter.

Stiegler became a philosopher through his reading of Plato's writings. His reminiscence took place in the element of the interpretation of texts that transmit the spirit of Plato. If he could arrive at the significance of the dialogues, that is because he was able to build 'a sustained relation'[26] to them through which he could individuate himself. That he imposed upon himself and systematically practised his routines and disciplines, reflects indeed how *anamnēsis* must compose with *hypomnēsis*. This is, in brief, the position Stiegler adopts in 'How I Became a Philosopher': we may all be devoted in potential to philosophy, yet to become a philosopher 'through acting',[27] we have to rely on a technical already-there that exceeds the mind to include it in vast mnemonic fields.

'Signifying practices that can rip'

Stiegler discovered the question of reminiscence as the question of the signifying practices that he was able to frequent. These practices are 'the supports of making-world'.[28] As a locality that gives place to intellective life, this world is irreducible in being irreducibly deictic. Whenever the 'practices in signifying fields' come to 'weaken, rip, and decay',[29] though, this world will start to lose its consistency. Locality then no longer gives place to *noēsis* and neither to noetic souls. Stiegler witnessed this possibility of decay in prison, and he constantly felt it as the danger facing him, 'at the limit'.[30]

As the living spirit is inhabited by the essential exteriority of signifying materials and practices, anamnesic forms of life are exposed from within to 'in- and a-significance'.[31] How could we articulate the decay and yielding that may thus affect the life of spirit, and the intellective practice of the philosophical discipline in particular?

Anamnēsis may lead to *dianoia*, to thinking with oneself. As it consists in the interiorization of dialogue and of the hypomnesic retentions that make for the long circuits of transindividuation, it involves the articulation of psychosomatic organs (the brain, for instance), social organs (the circuits of exchange, for instance) and technical organs (such as texts). Precisely this organological dimension of *anamnēsis* raises a pharmacological question, the elaboration of which defines Stiegler's organo-pharmacological approach. Essential to this approach is the reference to Gilbert Simondon, who considers that the explanation of psychic individuation demands the explanation of collective individuation and of transindividuation, defined as the individuation of what ties the *I* and the *we*: language, law, philosophy and all that 'constitutes for us a pre-individual fund'.[32]

That *anamnēsis* is sustained organologically, means first of all that knowledge is a *social* product, and not just a mental product.[33] An individual intelligence is never just individual. If it is the hypomnesic milieu that constitutes living memory as *knowledge*,[34] then one cannot ever think all by oneself. For the individual, the intermediary of *hypomnēsis* is 'the fact of being an element of a much vaster individuation'.[35] Technical objects are traces that connect psychic and collective individuation. To say that the retentional milieu that gives place to *anamnēsis* is pharmacological is to say that it is not strictly beneficial to noetic life, but only to the extent that it simultaneously wounds it. The milieu is remedy and cure at the same time. This very duplicity needs articulation in terms of both psychic and collective processes of individuation.

As a collection of archived engrams, knowledge retains the inscriptions of the old whence it comes and which it renews through *anamnēsis*. The conservation of these traces enables the constitution of circuits of collective individuation across time within the framework of a discipline[36] governing the connection between the minds which individuate in concert – the *I* and the *we* accessing their alterity and future through each other[37] – and in the course of an intergenerational transmission through which a process of transindividuation is concretized that produces the transindividual and makes for the significations that tie the *I*s and the *we*.

In this way, thinking and intelligence are always already collective. For sure, thinking is thinking for oneself; yet, insofar as it circulates, thinking surpasses itself, being always '*for* the other' and '*through* the other', 'through the other as thinker'.[38]

Technical individuation inevitably exceeds psychosocial co-individuation. In this excess lies the pharmacological character of the milieu. The irreducible disjunction between the technical milieu and the social systems makes the *pharmakon* of technical memory the source of both the intensification of the co-individuation of the *I*s and the *we*, and the obstruction thereof. It follows that the dynamism of the technical system requires procedures that allow the psychosocial co-individuation to partake in the individuation of 'its' technical milieu and to develop the necessary pharmacological ingenuity to adopt it, even if its toxicity is irreducible.

Philosophy's subject

Spending five years in passing to the limits of phenomenology, Stiegler refashions philosophy in terms of a reminiscence that is exposed to loss as it demands frequentation and requires access to the otherness of retentional inscriptions and of the others mediating access thereto. With the same stroke he recasts its subject. The spirit that achieves transcendence[39] on 'forgetting' its reliance on supplementation gives way to a soul whose dialectic life *just is* the reflexive relationship it entertains with essentially exterior technical retentions. Embattled over intelligence, this psyche never quite gets its act together. It is always somewhat lost amidst its concretizations,[40] to find itself at the receiving end of acting. In being moved by its materializations, it never ceases to divide and find itself other than itself.[41] If its *noēsis* cannot be traced back to some authenticating origin,[42] it also remains structurally incapable of completion.[43]

As a battle over psychic and collective co-individuation according to the conditions of technical memory, spirit certainly 'requires intelligence of technology, and intelligence about intelligence in its connection to technology'.[44] Yet, as technical materializations do not record matters that would have occurred *prior to* their exteriorization, spirit can never be supposed to be in control of 'its' concretizations. Since supplementary inscriptions partake in its individuation, it is largely invented by technical otherness. As such, it is promised an iteration that may renew it, albeit always somewhat accidentally and out of recognition. While 'things come to pass and *events happen*' in the relationship it maintains with technically organized matter, noetic co-individuation may hence truly 'constitute histories'[45] in which trajectories of *becoming* bifurcate and new pathways for transindividuation prime. Still, as iteration provides no guarantees, technical reinvention may always go wrong and short-circuit co- and transindividuation. In a way it always does go awry, even, since it catches spirit unawares, as its *default* of origin.

Engaging the collective

What Ernest Renan referred to as 'the Greek miracle' stands for the emergence, in the eighth century BC, of a new psyche and a new *polis*. In the wake of the invention of alphabetic writing a profound intellective change takes place that allows for the possibility to individuate oneself as a citizen. This ideal withers during the fifth century BC, when the sophists seize hold of writing to produce a discourse that is not meant to be true but to have effects. It is from this crisis of the Athenian democracy that philosophy arises. Philosophy is consequently always political: not so much because it is inaugurated as a critical reflection on the living conditions of citizens in the *polis* as a framework of *becoming*, but because of its engagement in a battle over intelligence, whose very instruments are 'weapons in a war for minds'.[46] Philosophy is from the start biased in favour of a structurally hypomnesic, hence fragile dialectic co-individuation, and it is this partisan slant that entails that 'the question of philosophy is first of all that of action'[47] and that the question of action is to be explicated in terms of a dressing of wounds, a 'pansement' of the sores of pharmacological supplementation. This question is about the appropriation of forms of knowledge, and clearly delineates the stakes of philosophy's defining battle with sophists: how to keep the co- and transindividuation open for future transformation?

The life and death of the proto-philosopher Socrates are of course exemplary in this connection. Philosophy gets going as a singular act of individuation that 'engages the collective to which the philosopher belongs in a co-constitutive individuation',[48] thereby 'inaugurating a new attitude, which is philosophy through acting'.[49] Set about as the discourse of an individuation, to philosophize is *eo ipso* to join in the *becoming* of organological *beings by default* who co-individuate what connects them *as per* the pharmacological conditions of noetic experience and knowledge.

It is of consequence that Socrates ties his *death* to the singular plurality of the *polis*, and not just his *life*. In facing the judgement of the city right up to the hemlock, Socrates bequeaths his passing as an obligation 'of continuing to interpret the laws of the City *beyond* his death, just as much as *from* that death'.[50] Stiegler considers it Socrates's 'genius' to tie us in this way to an *anamnēsis* of the city's constitution that ceaselessly defers its completion and is 'never to be *realized*'.[51]

That Socrates's death remains unfinished is critical to philosophy's acting as a discursive practice that unfolds through the singularity of an individuality, and 'as *unachievable*'.[52] It is precisely in being charged with the onus to keep its interpretation going, that his death intimates how it is not possible to constatively *objectify* what individuation *is*. This is why Socrates's thought is a nonknowledge:[53] in tying us to a structurally incomplete *anamnēsis* of the city's administration of the law, Socrates binds us to a reminiscence of the *default of being* that reaffirms the *to-come* of in-completion.

To articulate this default as a nonknowledge, we might say that incompleteness amounts to *not yet* knowing – 'or [not] any longer'[54] – how to live in a temporal locality. Time and again Socrates invites his interlocutors to reminisce their default of knowledge in relation to a life situation, thus clearing the horizon of their *becoming*. In having them partake of the default qua nonknowledge, he encourages them to pursue an individuation that is always at the same time a co- and transindividuation.

'Making (the) difference'[55]

To become a philosopher taking on the question of reminiscence, that is, of the frequenting of signifying practices that are the supports of making-world, is to assume a responsibility 'to the death' to interpret the laws of the *polis* as a framework of *becoming*. As he narrates his turn towards philosophy, Stiegler praises the laws of the republic 'that meant that there was a library in prison', and calls right away to rejuvenate 'their spirit through *anamnesis*'.[56] He would later contribute to such a renewal by recasting the book house as a 'collection of potentialities awaiting their reading so as to be actualized, noetically singularising life as the neganthropy' that is 'constituted by the *anamnesis* of neganthropic predecessors'.[57] This layered description of the athenaeum – as an accumulation of *signifying materials* conductive to the transformative actualization of new significations, and as an archive of prevenient retentions (not to say technical protentions)[58] engaging *noetic singularizations-to-come* – serves to incite the organizing authorities to give consideration to the negentropy that defines the specifically organological form of life. As public powers adopt this *neganthropy* as a point of attention, policymaking may come to reflect present concerns for the evolution and diversification of noetic manners of life.

Socrates faces similar concerns, as sophists venture to objectify and control the co-individuation of Athenians through writing in an effort to substitute *dianoia* and dianoetic singularization with the clichés of a ready-made thinking. Socrates is on a mission to transform writing into the technics of an individual and collective intelligence that intends to form a social (political) apparatus. Through an attentional psycho-technique called maieutic and an interiorization that is the product of a *hypomnēsis* (cf. *Meno*), Socrates's organology is as much a political organology as an organology of knowledge. It is this double dimension, political and epistemological, that constitutes the *we* that is the system of care and attention called the *polis*, created through a battle for intelligence.[59] In the fight he puts up, Socrates takes care of *pharmaka* through their careful use against the perverse effects of *pharmaka*. Antagonistic to the sophists, he takes charge of the social through the imposition of a *de jure* psychopolitics upon a *de facto* psychopower.[60]

Now, if there is a thread running through all of Stiegler's interventions and performances, writerly or other, from the beginning to the end,[61] it is the incitation to make the difference, or *différance* between fact and law, facts and rights. That he time after time called to make and cultivate this difference, is the signature stamp of his *theoria* qua *praxis*, his '*penser/panser*'.[62]

That this difference is to be *made*, means first and foremost that it cannot be *observed*. The law that the difference makes is not *given*. As it has to be *giving*, it might be explicated as a right to singularity.[63] While a singular relational being proves to be an ingenious improvisator affirming the default of in-completion as the resource and potential for an exceptional *becoming*, the law of singularity may be said to be *giving* by definition, opening out from the confines of the given on to the *to-come* of improbable bifurcations. As Stiegler puts it pointedly, 'l'idiome fait la différence':[64] what makes (the) *différance* is, if not the idiom that 'technics is' and that has 'always already been [set in motion /undermined] [*entamer*] by a default', then the idiomatic invention of singularity.[65] Note that *différance* is here to be defined in terms of technics as time, that is, of the incalculable *to-come* that the already-there holds in reserve: 'I remained temporal in remaining idiomatic, and thus I never finished becoming'.[66] The obliging insistence that *différance* be made, organologically, with care and attention, definitely brings out the (co-)existential stakes of *Technics and Time*: the in-determination of technical forms of life, in 'a certain rapport with death'.[67]

'A philosopher in deeds'[68]

Socrates takes the techno-material and imperviously secret base of spirit's life into account. In betrayal of him, the later Plato comes to oppose *anamnēsis* to *hypomnēsis*, thus opening the transcendental horizon of the changeless ideas. In disowning technics, however, idealist philosophers are unable to move beyond edifying responses in reaction to new technical transformations – 'chatter', indeed.[69] A compositional mode of thinking, on the other hand, explains *noēsis* in terms of a history of technically organized matter. In recognition of the brittleness of signifying practices, and of the fragility of the psychosocial co-individuation in accordance with the conditions of organized matter, it comes to recommend a politics of spirit in its battle with its concretizations, and to press for a politics of the milieu that gives place to noetic life. In encouraging us to participate in the technical invention that invents us, this thinking style can indeed meet the ambition to instigate the individuation of the singular plurality to which the philosopher is tied.

So it comes that as a philosopher – engaging the search for pathways of *becoming* sustained by technics, and doing so for love of the *polis* as a framework thereof – Stiegler invariably challenges his interlocutors to gather intelligence about *noēsis* in its connection to technical individuation, and encourages them again and again to partake in organological experimentation, to dream up technical media and tinker with *pharmaka*. In so doing, he takes part in their becoming genuine amateurs of an *unachievable* dialogical *noēsis* that is inhabited by a principally unavailable outside. Serving the same end is the practical, concurrently political *and* epistemological work that he is involved in as a philosopher who takes *noēsis* back to what he finds at its root, technically organized matter, and that leads him – among other things – to write annotation-software fitting up beholders of audiovisual content to become the co-producers thereof, conceive human/machine interfaces as objects of desire granting an opportunity to sustain *practices* rather than *usages*,[70] consult teenagers in their adoption of media, design ways to constitute a museum public, stylize citizens' participation in drawing up party-programmes, foster contributory research methodologies, incite students to practice individuating 'journeys of knowledge'[71] or work out the forms of polity and organization global crises such as Covid-19, climate destruction and migration demand.

'I remain a materialist'[72]

In the decade prior to his incarceration, Stiegler 'had no philosophy': being a combative materialist, he believed that philosophers were 'necessarily on the wrong side' and that 'true philosophy, that is, *philosophy in action(s)*, was politics'.[73] On becoming a philosopher, he could no longer identify as a materialist in this sense. In recognizing what idealist philosophy disavows – that philosophy as *anamnēsis* requires a *hypomnēsis*, and that as a fabric of signifying practices it demands techniques of co-individuation – he remained a materialist, though. Obviously, the rearticulation of matter as a history of technically organized matter in relation to the human, brought about a change of focus[74] away from property relations to patterns of organized 'loss of individuation'.[75] This shift allowed him to broach a new source of potentials and dynamics[76] – namely, technical alterity – and to battle for a 'politics of the passage to the philosophical act'.[77] In recognizing the constitutive role of technics, he was in a position to reflect on contemporary technological supplementation beyond the terms of destitution and calculation. Indeed, if it is explicitly as technical consciousness that noetic souls invent themselves, it follows that experimentation is proper and vital to them. Hence Stiegler's reiterated call not to withstand new transformations and critique them negatively, but to critique them positively and to reinvent, both in terms of an intelligence of the hypomnesic bases and of a future for anamnesic individuation.[78]

'It is a matter not of resisting but of inventing'[79]

Faced with the toxicity of a *pharmakon* – and of digital reticulation in particular, leading to a permanent and generalized connectivity that takes hold of the memories and anticipations which constitute the *Is* and the *we* – it cannot be considered adequate to *resist* these developments: 'it is necessary to transform them'.[80]

Speaking in 2016 – in the name of Ars Industrialis, *Pharmakon.fr*, IRI, the programme running in Plaine Commune, and the Chair of Contributory Research endowed at the Maison des sciences de l'homme Paris Nord – Stiegler spells out what the requisite transformation should include

1. the implementation of macro-political concepts and methods of a new kind, and the reinvention of forms of public power that may challenge libertarian ultraliberalism;
2. the creation of the means to develop new conceptions of theoretical information science and to prescribe new data architectures so as to counter the interiorization of a state of fact enforced by the industrial technologies of the data economy and resulting in the dissolution of the web's promising potentials in the correlationnist calculus of big data that valorizes averages and liquidates exceptions statistically.[81]

It is important to notice that the recently established Internation Collective (2018) and Association of the Friends of the Thunberg Generation (2019) are as primarily involved in the reinvention of forms of co-individuation, public power and technological practice, as are the aforesaid institutions, although they are first defined in reference to an ecological crisis that is also a crisis of the intergenerational relationship. It is telling, in this regard, that the Internation Collective advocates a decarbonization that passes through a deproletarianization,[82] defined as the struggle and politics to reclaim the knowledge we have lost through the organized proletarianization of our lives. The transformation of the systems that affect psychic and collective motivation is judged to be a prerequisite for the transformation of the economy.[83] Accordingly, the reinvention of a new organology based on the potentials contained in the digital technical system enjoys priority.[84] Indeed, to 'restart the world system'[85] we must first take back our noetic co-individuation, that is, our specifically technical form of life.

'Personally, I only work with collectives'[86]

Over the past years, Stiegler has been keen to point out that he was a team player, working with others on specific problems, contributing to projects in a context of sharing and generosity that is beneficial to intellective renewal. The motivation he provided was twofold. To begin, given the urgency and complexity of 'the hyper-crisis'[87] hitting the planet, one simply cannot do all the research by oneself. More important, though, is the observation that science today no longer thinks carefully: '*la science ne panse pas*'.[88] Over the past century, scientific knowledge has undergone a transformation. As a product of the understanding, it is certainly well-constructed and supported by data. However, since it does not produce any purposiveness, it cannot really qualify as knowledge. Knowledge proper finds its apogee in the projection of an interpretative decision.[89] It does not reinforce what is given, and goes beyond that which intensive computation rolls out as 'the real'. In displaying creativity, it facilitates the occurrence of bifurcations through which neganthropic possibilities arise.[90] It thus fosters the inscription of 'openings' in the process of becoming, so that '*le devenir*' does not turn into a totally immanent affair – in which 'the past is eating the future'[91] – and may instead be pursued as a process in which everything is still to be decided, to begin with 'the time of [co-]individuation'.[92]

If technoscience no longer cares thoughtfully, that is because it has been taylorized through an ever-increased specialization and has become a science of parts to which there is no whole.[93] Whence Stiegler's plea to reconstruct knowledge through the reconstitution of a holistic viewpoint that allows for the perception of elements as parts of a whole that is 'superior to the sum of its parts'.[94] In being projected 'onto the synthetic plane' by 'the faculty of reason',[95] which exceeds the 'calculation and analysis'[96] provided by the understanding, such a whole relates to the very possibility of the future: as it creates that which is 'real', and with it, possibility, it generates transformations that prior to its invention were not even possible.

Now the establishment of contributory research groups is precisely to advance work on concrete issues in view of the elaboration of such types of knowledge beyond specialization that 'convoke reason'.[97] These forms of knowledge discover in the *pharmakon* the dynamics of a co-existence that knows how to singularize itself and become the exception to the rule of adaptation that wounds all life, thus initiating a culture that poses 'in principle' what would be 'the principle of a political economy understood as a will to believe', namely, 'that the development of the technological process must be the development of singularity'.[98, 99]

Notes

1. Bernard Stiegler, 'How I Became a Philosopher', in Bernard Stiegler, *Acting Out*, trans. David Barison, Daniel Ross & Patrick Crogan, Stanford: Stanford University Press, 2009, 1–35, 35.
2. Stiegler, *Acting Out*, 24; 20.
3. Stiegler, *Acting Out*, 3.
4. Stiegler, *Acting Out*, 29–30.
5. Stiegler, *Acting Out*, 14; 'give place to' translates 'donnent lieu à' (give rise to, generate).
6. Renewal assumes the creation of an opening for alteration. This is not without risk, since iteration – no less than the future – involves expropriation and implies a relation to radical displacement, for which the paradigmatic reference has to be death. So it comes that the textualities that move philosophical *noēsis* in particular – 'its' texts and syntaxes, circuits and institutions, scenes and histories; 'its' relationships to the arts and sciences, to politics (Bernard Stiegler, 'Le concept d'"Idiotexte": esquisses', *Intellectica. Revue de l'Association pour la Recherche Cognitive* 53–54/1–2 (2010): 51–65, 52, available at: https://www.persee.fr/doc/intel_0769-4113_2010_num_53_1_1178) – open it uncontrollably to 'in- and a-significance' (Stiegler, *Acting Out*, 26). It is precisely this *non-closure* at 'their' limits – the impossible *give* at the core of signifying materials and practices putting their consistence on the line – that creates the space and chance for their unexpected renewal-as-reinvention by technical alterity.
7. Stiegler, *Acting Out*, 1; 20; 12.
8. Stiegler, *Acting Out*, 3.
9. Stiegler, *Acting Out*, 15.
10. Bernard Stiegler, *States of Shock: Stupidity and Knowledge in the 21th Century*, trans. Daniel Ross, Cambridge: Polity, 2015, 187: 'the *hypomnesic* overflow both frames academic life […] and at the same time constitutes its heteronomy, because this also frames its outside'.
11. Stiegler, *Acting Out*, 18.
12. Stiegler, *Acting Out*, 18–19.
13. Stiegler, *Acting Out*, 26.
14. Stiegler, *Acting Out*, 11–12.
15. Stiegler, *Acting Out*, 23.
16. Stiegler, 'Le concept d'"Idiotexte": esquisses', 62.
17. Bernard Stiegler, *Technics and Time, 2: Disorientation*, trans. Stephen Barker, Stanford: Stanford University Press, 2009, 65.
18. Stiegler, *Acting Out*, 30.
19. Stiegler, *Acting Out*, 20.
20. Stiegler, *Acting Out*, 28.
21. Stiegler, *Acting Out*, 16.
22. Stiegler, *Acting Out*, 29.

23 Stiegler, *Acting Out*, 22.
24 Bernard Stiegler, *Philosophising by Accident: Interviews with Élie During*, ed. and trans. Benoît Dillet, Edinburgh: Edinburgh University Press, 55.
25 Stiegler, *Philosophising by Accident*, 55.
26 Stiegler, *Acting Out*, 28.
27 Stiegler, *Acting Out*, 2.
28 Stiegler, *Acting Out*, 29.
29 Stiegler, *Acting Out*, 29.
30 Stiegler, *Acting Out*, 29.
31 Stiegler, *Acting Out*, 26.
32 Stiegler, *Acting Out*, 4–5.
33 Bernard Stiegler, 'Lezende hersenen, digitale hersenen en algemene organologie', vertaling Jelle Zeedam, *Ethische perspectieven* 24/3 (2014): 215–29, 216.
34 Stiegler makes a distinction between knowledge and cognition. Since knowledge depends on sociality and technicity, it transforms cognition. Cf. Terence Blake, 'Bernard Stiegler, knowledge, cognition', *Agent Swarm* (24 January 2014), available at: https://terenceblake.wordpress.com/2014/01/21/bernard-stiegler-cognition-knowledge/.
35 Stiegler, *Acting Out*, 5.
36 All knowledge is practised and developed through circuits of co- and transindividuation across time. These circuits are generally not constituted in the frame of a (school) discipline. It is precisely in thematizing the constitution of its circuits that a knowledge practice may attain the level of reflexivity considered typical of a discipline. What is commonly 'forgotten' in such thematizations, however, is the way in which the constitution of individuation circuits is enabled by technical memory. It thus stays out of view how the processes of co- and transindividuation through which a discipline is constituted are over-determined by the characteristics of the archival supports.
37 Stiegler, *Acting Out*, 30.
38 Bernard Stiegler, *Taking Care of Youth and the Generations*, trans. Stephen Barker, Stanford: Stanford University Press, 2010, 34.
39 Stiegler, *Acting Out*, 29.
40 Bernard Stiegler, *Technics and Time, 3: Cinematic Time and the Question of Malaise*, trans. Stephen Barker, Stanford: Stanford University Press, 2011, 49; Bernard Stiegler, *De la misère symbolique 2: La catastrophè du sensible*, Paris: Galilée, 2005, 93.
41 Stiegler, *Acting Out*, 4.
42 Stiegler, 'Le concept d'"Idiotexte": esquisses', 63.
43 Stiegler, *Acting Out*, 16.
44 Stiegler, *Taking Care of Youth and the Generations*, 34.
45 Stiegler, *Acting Out*, 4.
46 Stiegler, *Taking Care of Youth and the Generations*, 33.
47 Stiegler, *Acting Out*, 7.
48 Patrick Crogan, 'Bernard Stiegler: Philosophy, Technics, and Activism', *Cultural Politics* 6/2 (2010): 133–56, 135.
49 Stiegler, *Acting Out*, 6.
50 Stiegler, *Acting Out*, 6.
51 Stiegler, *Acting Out*, 4.
52 Stiegler, *Acting Out*, 6.
53 Stiegler, *Acting Out*, 6.

54 Jacques Derrida, *Specters of Marx: The State of the Debt, the Work of Mourning and the New International*, trans. Peggy Kamuf, London: Routledge, 1994, 65.
55 Stiegler, *Technics and Time, 3*, 157.
56 Stiegler, *Acting Out*, 23.
57 Bernard Stiegler, *The Age of Disruption: Technology and Madness in Computational Capitalism*, followed by *A Conversation about Christianity with Alain Jugnon, Jean-Luc Nancy and Bernard Stiegler*, trans. Daniel Ross, Cambridge: Polity, 2019, 27.
58 Stiegler, *States of Shock*, 144.
59 Stiegler, *Taking Care of Youth and the Generations*, 65; the *polis* is to form a 'community of the *default* of community' (Bernard Stiegler, *Technics and Time, 1: The Fault of Epimetheus*, trans. Richard Beardsworth and George Collins, Stanford: Stanford University Press, 1998, 193).
60 Stiegler, *Taking Care of Youth and the Generations*, 35; the translation suggests – misleadingly, I believe – that it is the 'de facto psychopower' that 'imposes a psychopolitics of law' (see Bernard Stiegler, *Prendre soin 1: De la jeunesse et des générations*, Paris: Flammarion, 2008, 68: 'un psychopouvoir économique de fait auquel il faut imposer une psychopolitique de droit').
61 Stiegler, 'Technologies de la mémoire et de l'imagination', *Réseaux* 16/4 (1986): 61–87, 65; Bernard Stiegler, 'What Is Called Caring? Thinking beyond the Anthropocene', in Bernard Stiegler, *The Neganthropocene*, ed., trans., and with an introduction by Daniel Ross, London: Open Humanities Press, 188–270, 250.
62 The following line from *Acting Out* (6–7) might well have served as an epigraph to this essay: 'Philosophical *saying* is necessarily *also* a *doing, to the death*, and this *theoria* is always also a *praxis* – failing which it is nothing but chatter'.
63 Stiegler, 'Le concept d'"Idiotexte": esquisses', 63.
64 Stiegler, 'Le concept d'"Idiotexte": esquisses', 63.
65 Stiegler, *Technics and Time, 2*, 156; 149 (trans. modified). Stiegler considers the (singularity of the) idiom to be 'recurrent': 'always already effaced', it 'remains ineffaceable, remains ineffaceably, because it is the law of the remainder: of the accident, of idiocy' (Stiegler, *Technics and Time, 2*, 84).
66 Stiegler, *Acting Out*, 25.
67 Stiegler, *Technics and Time, 2*, 157; Bart A. Buseyne, 'Onbepaald door de techniek', *De Uil van Minerva* 28/4 (2015): 319–34. In a discussion with Aristotle on the (community of the) senses, Stiegler notes how the *logos* can only mark the difference between the senses 'while losing it' (Stiegler, 'Wanting to Believe: In the Hands of the Intellect', in Bernard Stiegler, Bernard Stiegler, *The Decadence of Industrial Democracies. Disbelief and Discredit, Volume 1*, trans. Daniel Ross & Suzanne Arnold, Cambridge: Polity Press, 2011, 131–62, 181 fn. 7, translation modified); we take it that the difference between fact and law can similarly only be marked 'while losing it': it appears in disappearing, not unlike the mortal and *idios*.
68 *Plutarch's Morals*, trans. from the Greek by several hands, 14th ed., London: Braddyl, 1704, 450.
69 Stiegler, *Acting Out*, 7; cf. n. 62.
70 Bernard Stiegler, 'Pleasure, desire and complicity' (October 2015, Mexico City), available at: https://www.academia.edu/12693814/Bernard_Stiegler_Pleasure_Desire_and_Complicity_2015_.
71 Bernard Stiegler, *Nanjing Lectures 2016–2019*, ed. and trans. Daniel Ross, London: Open Humanities Press, 2020, 92.
72 Stiegler, *Acting Out*, 32.

73 Stiegler, *Acting Out*, 31.
74 Anne Alombert, 'Du prolétariat à la prolétarisation: vers une nouvelle critique de l'économie politique dans le contexte de l'automatisation numérique', *Implications philosophiques* (19 October 2018), available at: http://www.implications-philosophiques.org/actualite/une/du-proletariat-a-la-proletarisation/.
75 Stiegler, *Acting Out*, 32.
76 Stiegler, *Acting Out*, 17.
77 Stiegler, *Acting Out*, 22.
78 Stiegler, *Taking Care of Youth and the Generations*, 70.
79 Bernard Stiegler, 'La prison a été ma grande maîtresse', *Philomagazine* (27 September 2012), available at: https://www.philomag.com/articles/bernard-stiegler-la-prison-ete-ma-grande-maitresse.
80 Bernard Stiegler, 'Postface', in Ippolita, *Internet: l'illusion démocratique*, traduit de l'espagnol par Vivien García, postface de Bernard Stiegler, Paris: La Différence, 2016, 158–70, 164.
81 Stiegler, 'Postface', 164–5.
82 Bernard Stiegler & Collectif Internation, *Bifurquer: 'Il n'y a pas d'alternative'*, précédé d'une lettre de Jean-Marie Gustave Le Clézio, suivi d'une postface de Alain Supiot, Paris: Les liens qui libèrent, 2020, 33; Bernard Stiegler & The Internation Collective (ed.), *Bifurcate: 'There Is No Alternative'*, preceded by a letter from Jean-Marie Le Clézio, with an afterword by Alain Supiot, and a lexicon by Anne Alombert and Michal Krzykawski, trans. Daniel Ross, London: Open Humanities Press, 2021, available at: http://www.openhumanitiespress.org/books/titles/bifurcate/, 26.
83 Stiegler & Internation, *Bifurquer*, 220; Stiegler & Internation, *Bifurcate*, 164.
84 Bernard Stiegler, *Automatic Society, Volume 1: The Future of Work*, trans. D. Ross, Cambridge: Polity, 2016, 139.
85 Bernard Stiegler, 'Restarting the World System', in Anders Dunkers, *Rediscovering Earth: Ten Dialogues on the Future of Nature*, London: Or Books, 2021, 93–117.
86 Bernard Stiegler, 'Même s'ils le voulaient, les Etats n'auraient pas les concepts pour changer', *Libération* (20 March 2020), available at: https://www.liberation.fr/debats/2020/03/08/bernard-stiegler-meme-s-ils-le-voulaient-les-etats-n-auraient-pas-les-concepts-pour-changer_1780988/.
87 Bernard Stiegler, 'Démesure, promesses, compromis 2. Incertitude et indétermination', *Mediapart* (7 September 2020), available at: https://blogs.mediapart.fr/edition/les-invites-de-mediapart/article/070920/demesure-promesses-compromis-23-par-bernard-stiegler.
88 Stiegler, 'Même s'ils le voulaient, les Etats n'auraient pas les concepts pour changer'.
89 Stiegler, 'Même s'ils le voulaient, les Etats n'auraient pas les concepts pour changer'.
90 Bernard Stiegler, 'Toute technologie est porteuse du pire autant que du meilleur', *Le temps* (22 March 2018), available at: https://www.letemps.ch/opinions/bernard-stiegler-toute-technologie-porteuse-pire-autant-meilleur; in choosing between possibilities opened up by technical retentions, a decision creates a bifurcation.
91 Timothy Morton, 'Essay by Timothy Morton' (s.d.), available at: https://landmarks.utexas.edu/content/essay-timothy-morton.
92 Stiegler, *Technics and Time, 3*, 102 (translation modified).
93 Stiegler, 'Même s'ils le voulaient, les Etats n'auraient pas les concepts pour changer'; cf. Stiegler, *Technics and time, 3*, 151–3, and Stiegler, *De la misère symbolique 2*, 93.
94 Bernard Stiegler, 'Five Theses after Schmitt and Bratton', in Stiegler, *The Neganthropocene*, 129–38, 135.

95 Stiegler, 'What Is Called Caring? Thinking beyond the Anthropocene', in Stiegler, *The Neganthropocene*, 188–270, 234.
96 Stiegler, 'What Is Called Caring? Thinking beyond the Anthropocene', 239.
97 Stiegler, 'What Is Called Caring? Thinking beyond the Anthropocene', 239.
98 Stiegler, 'Wanting to Belief. In the Hands of the Intellect', in Bernard Stiegler, *The Decadence of Industrial Democracies. Disbelief and Discredit, Volume 1*, trans. Daniel Ross & Suzanne Arnold, Cambridge: Polity Press, 2011, 131–62, 141.
99 I thank Judith Wambacq for recollecting that line in *Passer à l'acte* that explicates philosophical *theoria* as *also* a *praxis*; Pieter Lemmens, Georgios Tsagdis, and Paul Willemarck for their percipient comments.

Chapter 12

TAKING CARE OF DIGITAL TECHNOLOGIES

Vincent Puig

To think technics in relation to time and spirit is the unfinished work Bernard Stiegler has left for us to do.[1] For generations to come, his oeuvre will be an infinite source of questioning, that is, of knowing, while it is precisely through its temporality and multiple historicity that technics can be said to constitute the 'milieu of knowledges' (*milieu des savoirs*).[2]

While he approached the digital as such an elementary milieu, Stiegler considered that he first had to study its *organology* in general terms.[3] As he qualified himself along the way, he could eventually enter into the specifics of its pharmacology, bearing in mind that any organ – whether biological, technical or social – can be a poison or a remedy, depending on whether we take care of it, or not. I would like to demonstrate, with specific reference to some of the projects in which I collaborated with him, that Stiegler knew well how to attend to what he called 'technologies of spirit' (*technologies de l'esprit*).[4] This showed, first of all, in the way he experienced these technologies concretely through the techno-cultural projects and development programmes he set up, and in the manner he criticized their incompletion positively so as to take a view on technology's becoming (*à-venir*). It was also reflected in the way he took on digital technologies as objects of a collective desire, and did so from a pharmacological perspective, with due respect for the intermittence that occurs necessarily between the synchronic and the diachronic, the stereotypical and the traumatypical, the calculable and the incalculable. It manifested, finally, in his determination to articulate this very intermittence in the dominant field of today's 'automatic society', the economy: since all knowledge comes from a transindividual experience, he projected to give concrete form to a contributory economy between *ponos* (labour) and *ergon* (work), *negotium* and *otium*, employment and work.

Memory and writing, retentions and protentions, building the organology of spirit

As the originator of a 'general organology',[5] Stiegler knew well how to take care thoughtfully. His attention was based on an inexhaustible desire to gain insight into the very organs of this care and desire, to understand their conditions of possibility in the registers of both the biological and the sociopolitical, and to comprehend practically their organogenesis and potential to defer entropy.

At the Institute for Research and Innovation (IRI) he proved to have a real knack for developing the negentropic potential of these organs down to the smallest details of their technological or methodological implementation. A few weeks before his death, we had a lively discussion on the organology of the Zoom platform for video communication and web conferencing, on the political dimension of its data collection model, and especially on the issue of attention design. Mindful of the loss of attention Zoom provoked, he argued that messages should not be displayed on the screen. While advising against a general use of the chat during seminars, he would indicate how it might be used profitably by a few participants for the specific purpose of annotating and indexing the video recording, of adding keywords, bibliographic references and comments. In 2020, standing in front of members of the Thunberg generation – some of whom born in the twenty-first century– he delivered a most frank speech that struck with lightning force as he was reactivating all the organological knowledge he had cultivated since the '80s of the previous century.

Grammatization of spirit

When he served in 1987 as the curator of the 'Mémoires du Futur' exhibition (1987) Stiegler introduced the general public to the history and future of what he then called hypomnesic supports, and would later refer to as tertiary retentions and exosomatic organs: the appearance of writing, the first libraries, printing, postal networks, telegraph, telephone, radio, television but also optical discs, audiovisual editing benches, video recorders, telex, news writing tools, search engines, etc. Shortly thereafter, in 1989, he led a research group – consisting of writers, poets, musicians, scientists and computer scientists – to develop a computer-assisted reading station (PLAO; Poste de lecture assistée par ordinateur) that anticipated the tools later developed at the Bibliothèque Nationale de France. Once he had formalized the annotation system that he had used for his thesis, and in particular for his research of Husserl's corpus, it could be modelled at the Université de technologie de Compiègne (UTC), before being implemented, in 1992, on a Sun computer.[6] If much of the ingenuity of these devices has, unfortunately, been lost, that is not only because of their technological obsolescence; it is due as much to the increasing entropy of digital technologies, built as they are on a notion of information for which there is *as of yet* no alternative, and which even quantum computing does not question: the notion of information as the inverse of entropy.

To take a closer look at on the characteristics of this annotation tool, we might refer to a text published at the time that describes the grammatization protocol it implements in terms of four types of operations:

1. hierarchy (the creation of unitary meaning, by underlining text fragments or drawing lines in the margins);
2. qualification (keywords, annotations, comments);

3. navigation & research (search windows, creation of links, categories of links specifying the nature of attached documents: canonical comments, translations, manuscripts, references used by the commented text, etcetera);
4. representation (various graphs, 'views' on the text corpus ...).[7]

The tool did not propose any a priori categories for annotation, as it was to be adjustable to many usages, in accord with the Standard Generalised Markup Language (SGML).[8] The concern for genericity and normativity that underlies its design comes into tension with the requirements of a computability that is to be supported only by a mode of statistical data analysis implementing an automated categorization that does not involve any interpretation, naming or standardization in terms of content.[9] This is why research and innovation at IRI has always been oriented towards the development of categorization tools that may be configured by the members themselves of a working group, and that can be adjusted to the objectives they set for themselves.

The constant conflict one encounters in Stiegler's work between organological considerations on the one hand, and philosophical trains of thought on the other, is somewhat analogous to the way in which Simondon worked out his theory of individuation starting from his electro-mechanical tinkering. Except that for Stiegler, technics clearly counted as the basis of what he called – with Simondon – transindividuation.

This tension was at the centre of further developments made in the 90s:

– first at the COSTECH laboratory[10] that was founded by Stiegler at UTC in 1993, and that launched the Digital territories project[11] introducing the highly innovative concept of the pre-indexed video form;
– then at the Institut national de l'audiovisuel (INA; English: National Audiovisual Institute), of which he was deputy director from 1996 to 1999;
– and from 2001 till 2006 at the Institut de recherche et coordination acoustique/musique (IRCAM; English: Institute for Research and Coordination Acoustics/Music), where I would set up with him a hypermedia studio similar to the one he had created at INA with Jean-Pierre Mabille and Xavier Lemarchand.

It is at the latter institution that the annotation software was developed, thanks to the help of Nicolas Donin, for a computer-assisted way of listening to music. Musicologists were then invited to use the programme to document their interpretation of musical performances, and to create so-called *Signed Listenings*. The tool was later reworked at Centre Pompidou and at IRI, and used to produce *Signed views*. Film critics were asked to elaborate their viewing of movies through the use of *Lignes de temps* (Timelines software). Further developments allowed the programme to be applied in real time within *Semantic HIFI*, adjusted specifically to the benefit of schools (*MusiqueLab Annotation*), and then largely promoted to amateur communities by the IRCAM Forum and the Resonances Festival of Technologies for Music.

Whether applicable to text, music or film, these technologies of spirit are all based on new forms of grammatization that articulate units of meaning as well as markings and modes of representation and navigation, and do so in the manner laid out in the article referenced earlier. Stiegler had an in-depth knowledge of documentary techniques that was not limited to archival science, or to what was later referred to as 'digital humanities'. He tried to fully explore the potential of digital grammatization to bring about new transductions (in Simondon's sense of the term) and to create innovative forms of negentropic organization that would produce new knowledge.

In an article from 2010, Stiegler explains how the 'idiotext', that is, the singular process of writing represented graphically by spirals of individuation, is always part of a dynamic of retentions and protentions, of memory and expectation that generates attention.[12] As 'a memory moved by its textuality',[13] the idiotext is sustained by a *différance* and 'an irreducible dispute [*différend*] that is technological in kind'.[14] For the technical dynamic to be 'reflexive', this *différend* and founding default (*défaut qu'il faut*) are to be carried on continuously. In many cases, Stiegler considered that this criterion was not met adequately. This prompted him to either abandon the device at hand or redevelop it towards a contributory and negentropic practice. In terms of design, reflexivity and *différance* of entropy are often sustained by processes of transduction that occur between reading and writing (this is why *Lignes de temps* does not separate a simple full screen reading from annotations), between analysis and synthesis (whence the articulation of *Lignes de temps* with mind maps, for instance), or between a spatialization process (i.e. the segmentation of a video) and a temporalization process (i.e. text-to-speech).

Reflexivity for transindividuation

In the way he organologically thinks through the design of the reflexivity that he sets as a requirement, Stiegler applies an understanding of libidinal economy.[15] The dynamic of the libido always manages, in one way or another, to articulate an automatic drive upon an infinite desire, an absolute synchronization to a video flow upon a diachronic writing space, phases of computable hypomnesic comprehension upon phases of anamnesic and incalculable disorder. Such articulations – which Stiegler situates somewhere 'between the dead and the living'[16] – also hold back entropy, albeit in a somewhat inadvertent way, namely temporal and memorial/historical processes that we might qualify today as anti-entropic.[17]

As an example of such experimental articulation, I may refer to the development of a set of contributory categorization protocols that favour transindividuation. The hermeneutic tool was implemented for the first time in a device for taking notes during a conference presentation. It was recast in the *Lignes de temps* software in 2009,[18] to be reworked again in 2010 into *Polemic Tweet*,[19] an application that is built on the Twitter Application Programming Interface (API).

The implement works in three stages:

1. a presentation of the protocol explains clearly to the contributors how their notes – or tweets – will be published in sync with the recording of the lecture they attend;
2. a live interface allows for the insertion, within tweets, of four so-called *metatags*, displayed by colours: comprehension (stereotypical retention marked in blue), trouble (traumatypical retention put in yellow), keywords (tinted in purple), free commentary (set in green);
3. the publication of the lecture recording, as indexed by the tweets, facilitates the identification of transindividuation points, keywords and/or references, and enables the use of an intra-video search engine that is based on the tweets.

We talk in terms of *metacategories* to indicate that the tags do not relate directly to the annotated content, but to the import of the contribution. This explains why the observed co-occurrences of metacategories over a circumscribed period of the lecture may disclose points of entry for further exchange and discussion between participants.

Taking care of intermittence: Towards an organology of benevolence

Stiegler has often related the concretization of noetic intermittence to Joseph Beuys's practice of social sculpture, to recast it later in life as a process of noetic gardening supported by the multiple retentions that constitute what he called the 'necromass'.[20] These figurations do not hint at a formal and stable design-project, since the concretization of intermittence must be metastable, in Simondon's words, and hence always remain open to bifurcation and negentropic organization, exactly as a garden should. Metastable digital systems are not entirely diachronic, they *play* the synchronic and *comprehend* the diachronic, the trouble and the surprise, what Stiegler calls a '*surpréhension*'.[21]

To illustrate this notion and experimental format of noetic intermittence, I often refer to the project *Penser-improviser* (English: to think-improvise) since it is undoubtedly in a discussion with the arts that Stiegler's thinking opened up the deepest bifurcations.[22] As part of the Mons 2015 European Capital of Culture festival, the Fabrique de Théâtre led by Valérie Cordy, and the Festival Les Inattendus co-organized musical improvisation workshops under the guidance by Bernard Lubat, which alternated with interventions by musicologists, writers and philosophers. As he had come to understand the authentic function of reason through his reading of Immanuel Kant and Alfred N. Whitehead, Stiegler considered that reason, even as it leaned on the understanding (*Verstand*; *entendement*), took an active part in the faculty of the imagination and was functionally operative in what he referred to as the faculty of collective, that is to say spiritual dreaming. The musicians, writers and stage actors participating in the workshops all testified how much they had to rely on deep automation if they were to move beyond in improvisation, and come to a true understanding of a text or piece of music. During the sessions, a large screen set up on the stage displayed a real-time transcription of the exchanges in the form of notes. As a computer was interacting with these exchanges, the transcript also presented an account of what was 'calculable' (predictable) and 'computable' by the processing unit.[23] This dialogue of processor and participants allowed for the isolation of phases of improvisation, that is, bifurcation in relation to the computer system. *Penser-improviser* thus explored the artistic boundaries between the synchronic and the diachronic, between the stereotypical and the traumatypical. It articulated live dialogue with a mode of musical data processing that did not only occur in real-time, since data processing simultaneously drew on the resource of a larger 'delay' via the interposition of the *Polemic Tweet* application annotating the musical categories as they were lighting up all along the interpretation and improvisation sessions and through the particularly juicy language of Lubat.[24] In thus being produced by the participants themselves, the annotations would serve as markers to facilitate a re-listening on stage, or on the website, with use of *Lignes de temps*.

This organology of the 'amateur' – relying as it does on multiple tertiary retentions and protentions – presupposes that we manage to regain control over the categorization processes that are today largely carried out in a purely computational manner by machines and through the production of statistical clusters of data.

Towards an artefactual benevolence?

In collaboration with the child psychiatrist Marie-Claire Bossière, Anne Alombert and Maël Montévil, Stiegler set up the Contributive clinic capacitation workshop in the county of Seine Saint-Denis. From 2017 onwards, its participants attempt to take care collectively of their young children who have been overexposed to screens. They do so within the framework of a contributory research group associating parents, caregivers and academic researchers.[25] Their care and attention are inspired by institutional psychotherapy, as introduced by François Tosquelles, and by the experience of Alcoholics Anonymous, as described by Gregory Bateson.[26] It rests on an analysis of the pharmacology of digital socio-technical devices that is carried out in a Winnicottian consideration of their potential for openness and desire, and of their prospective functioning as 'potential and transitional spaces'.[27] In line with Donald Winnicott and Robert Spaemann,[28] Emmanuel Benin has proposed the notion of *artefactual benevolence* (or *benevolent setting*) in his dissertation,[29] articulating it in terms of an experience of absence and lack that inspires trust, an 'openness and recognition of the other'[30] that may underpin the co-construction of 'artefacts and human relations' 'in a movement of mutual reinforcement'.[31] Belin cites the example of the house and its 'dwelling' (*l'habiter*) that we also find in Bachelard, Blanchot and Heidegger. It might be of interest to compare this conception with Stiegler's views on the importance of defect and *philia* as conditions of possibility of knowledge.

Even if Belin warns us against a hasty transposition of the notion of 'artefactual benevolence' into the theological field – this is a dimension in which Stiegler knew how to locate the *pharmakon* with unfailing accuracy – we can note that benevolence, in the Catholic tradition, is not in itself a (substantial) gift of Spirit, as it only consists in the (existential) fruits it bears, such as benignity and goodwill, first of all, or leniency and gentleness of the soul. Like knowledge, it is to be evaluated through its practice.

In the related, but different conception of spirit that we find in Simondon, the transindividual stage refers to a concurrently technical and spiritual relationship that exists between the individual and the group, and that transductively articulates affectivity and emotivity. Simondon takes the affective functions to be oriented according to bipolarities that are metastabilized in emotions: affections (happy/unhappy, for instance) have a meaning that is revealed by emotions (which are visible to the collective).[32] In a parallel way, the sensations linked to perception (hot/cold, light/dark) have direct consequences in terms of practice and action, as anchored in the collective.

The *Digital studies* programme that is developed at IRI and that looks into the articulation of calculable and incalculable functions, considers knowledge production to lean similarly on confidence in the technical artefact, in writing and publication, and hence on transindividuation. The programme sets out consequently to categorize the conditions of knowledge production in a way that does not come down to the construction of a relational ontology (which risks turning into a generalized computability à la Facebook) or to the literal calculation of emotions (emotional computing).[33] The Polemic Tweet device, for instance, arranges for the *delay* in the production of tweets (in real-time synchronic processes) through the application of a *tweet categorization protocol* that facilitates diachronic usages – during, but especially after the event – that invite the formation of groups of consensus and/or dissensus. The implement is hence not at all intended to signal emotions or to mark affections as vectors, functions or 'signs of becoming';[34] it rather means to articulate affection *and* action, the individual dimension of psychic development *and* the collective dimension of knowledge production.

A 'benevolent' social network designed for interpretative purposes should offer individuals the opportunity to project themselves in confidence; it should give groups the tools to deliberate and make collective decisions, and afford collaborators the possibility to editorialize the projects they are engaged in.[35] It also ought to provide partners the option to prioritize discussions according to their level of contribution. Exchange and sharing functionalities should, for instance, enable group members to mutually compare their work and to annotate one another's annotations. Yuk Hui and Harry Halpin have demonstrated how a social network like Facebook is built on the principle of Moreno's social graphs, that is, on the idea that the individual is the primary node in the network. To sidestep the techno-methodological individualism of the networking service – which tends to operate with an understanding of *profile* that no longer supports individuals, but creates 'dividuals' instead – Hui and Halpin introduced an approach to social relations that rests on the group in Simondon's sense of the term.[36] Coming first is then no longer the individual, but their relation to the 'associated milieu': their belonging to one or more group(s), their work on one or more project(s).[37] This focus on the group necessitates the design of another relation to calculation. Group-based networks do not run on algorithms that extract and process user data so as to steer behaviour; they operate with algorithms that are devised to recommend hermeneutical convergences and/or divergences, and to serve the automated suggestion of groups to be formed. A hermeneutical group should have the option of making decisions as to how the mutual sharing of contributions is organized, notably in terms of the network architecture (multicast and decentralized) and the data (open to deliberation). Is a group not characterized first and foremost by its autonomy, its ability to give itself rules and to practice normativity in Canguilhem's sense of the term?[38] The implication must be that the social network is indeed to be endowed with a governance rule, formalizable as in *Wikipedia* (although the latter gradually tends to automation). Since such a conception of the network is deeply attached to a dynamic that is both collaborative – in being based on common work – and contributory – in recognizing the contributors' contributions – it should put calculation to use for deliberative purposes, and not directly for predictive or performative ends.

Practice of knowledge and economy of contribution in the Covid-19 intermittence

With the launch of the 'Contributory learning territory program' in Plaine Commune, Stiegler and IRI gradually developed the model of a contributory economy that rests on yet another level of intermittence, with actors alternating between periods in which they develop knowledge and enjoy a contributory income, and periods in which they put this cultivated knowledge to practice through employment that conditions subsequent access to contributive income rights. This model is inspired by a specific form of intermittence, the regime for intermittent entertainment workers in France and Belgium. In the context of a tendential reduction of employment rates due to automation, this model requires us to invent forms of commons production that are similar to the ones developed in free software circles.[39] This intermittence hinges on a very concrete difference and articulation between employment and work, *ponos* and *ergon*, *negotium* and *otium*, or between labour and work. Following André Gorz, Antonella Corsani articulates the distinction in terms of heteronomy and autonomy.[40]

At a time in which the health crisis is blurring all the benchmarks that hitherto marked the difference between the private world and the world of work, we urgently need to mind and take care of intermittence, in matters of spirit as well as action, between our automatisms and our bifurcations, between the calculable and the incalculable, between desire and drive. The Covid-19 crisis exposes and reinforces 'cleavages' of all kinds, to endorse a new 'schize' whose pathological consequences still need to be assessed if we are to be able to devise and implement a regime of intermittence that can counter entropic synchronicity, to deploy what Antoinette Rouvroy calls 'heterochronicity'.[41]

The workshops conducted as part of the Contributory learning territory project are empowered by a constant concern to promote negentropic bifurcations. This care and attention are supported by a method to analyse the value that is the specific fruit of knowledge practices, and that is enriched as it is further qualified and labelled on a web-based platform. The platform's technology enables deliberation on the process of value creation and provides an innovative auditing programme for registering value. The accounting application is inspired by the CARE-TDL method that recognizes liabilities as a form of debt vis-à-vis local knowledge.[42] It is here – at the core of a system of capacitation that may further economic activity – that the organological foundations introduced by Stiegler continue to unfold.

Notes

1   In the introduction to the collected edition of La technique et le temps, Stiegler announced four forthcoming volumes: '4. L'épreuve de la vérité dans l'ère post-véridique, 5. Symboles et diaboles, 6. La guerre des esprits, 7. Le défaut qu'il faut. Idiome, idios, idiotie' (Stiegler, *La technique et le temps 1: La faute d'Épiméthée – 2: La désorientation – 3: Le temps du cinéma et la question du mal-être suivis de Le nouveau conflit des facultés et des fonctions dans l'Anthropocène*, Paris: Fayard, 2018, 12).
2   Translating this notion leads to a definition of *organologie des savoirs* that does not confuse *milieu* and *environment* and that makes a distinction between *savoir* and *connaissance*.
3   Bernard Stiegler (dir.), *Digital Studies: Organologie des savoirs et technologies de la connaissance*, Limoges: FYP éditions, 2014.
4   Better translated, I feel, as 'technologies of spirit' than as 'technologies of mind' which refers to cognitive sciences (cf. Marvin Minsky, *The Society of Mind*, New York, NY: Simon and Schuster, 1986).
5   Which Stiegler developed in particular through his readings of the work of Georges Canguilhem, André Leroi-Gourhan and Gilbert Simondon.
6   Developed by Berger-Levrault, a company that was directed at the time by François Chahuneau, and later changed its name to Diadeis.
7   Bernard Stiegler, 'Annotation, navigation, édition électronique: vers une géographie de la connaissance', Linx H-S 4 (1991): 121–31, available at: https://www.persee.fr/doc/linx_0246-8743_1991_hos_4_1_1191. Regarding grammatization, cf. Sylvain Auroux, *Les révolutions technologiques de la grammatisation*, Bruxelles: Mardaga, 1995.
8   SGML was the predecessor to HyperText Markup Language (HTML).
9   The concern for genericity and normativity is shared today by the World Wide Web Consortium (W3C), among others, at least as the Resource Description Framework (RDF) standard is concerned.
10 See the website of the *Connaissance Organisation et Systèmes Techniques* laboratory (English: Knowledge, Organisation, and Technical Systems), available at: http://www.costech.utc.fr/.
11 COSTECH collaborated to this effect with Olivier Landau at Orange's Sofrecom.
12 Bernard Stiegler, 'Le concept d'"Idiotexte": esquisses', *Intellectica. Revue de l'Association pour la Recherche Cognitive* 53–54/1–2 (2010): 51–65, available at: https://www.persee.fr/doc/intel_0769-4113_2010_num_53_1_1178.

13 Stiegler, 'Le concept d'"Idiotexte": esquisses', 52.
14 Stiegler, 'Le concept d'"Idiotexte": esquisses', 58.
15 Bernard Stiegler, 'Désir et connaissance: Le mort saisi par le vif. Éléments pour une organologie de la libido', *La Deleuziana. Online Journal of Philosophy* 6 (2017): 68–81, available at: http://www.ladeleuziana.org/wp-content/uploads/2017/12/Deleuziana6_68-81_Stiegler.pdf.
16 Bernard Stiegler, 'Désir et connaissance: Le mort saisi par le vif. Éléments pour une organologie de la libido', 69.
17 With reference to recent research and discussion, see Giuseppe Longo & Maël Montévil, 'The Inert vs. the Living State of Matter: Extended Criticality, Time Geometry, Anti-Entropy – an Overview', *Frontiers in Physiology* 3/39 (2012): 1–8, available at: https://www.ncbi.nlm.nih.gov/pmc/articles/PMC3286818/; DOI: https://doi.org/10.3389/fphys.2012.00039, and Anne Alombert & Michał Krzykawski, 'Vocabulaire de l'Internation', *Appareil* (3 February 2021), available at: http://journals.openedition.org/appareil/3752; DOI available at: https://doi.org/10.4000/appareil.3752.
18 Available at the IRI website, http://ldt.iri.centrepompidou.fr.
19 Availabe at Polemic Tweet website, http://polemictweet.com.
20 Bernard Stiegler, *Qu'appelle-t-on panser? 2: La leçon de Greta Thunberg*, Paris: Les Lies qui Libèrent, 2020, 17.
21 Bernard Stiegler, 'Shakespeare-to-peer', *Lemagazine* (25 August 2011), available at: https://archive-magazine.jeudepaume.org/2011/08/shakespeare-to-peer/index.html. A 'surprehension' goes beyond understanding and exceeds the understanding. It is that, by which one can begin to think 'for oneself' and exercise the synthetic powers of reason to judge what until then one tended to consume. Cf. Michal Krzykawski, 'L'idiodiversité, le sens et la raison de traduire. Hommage à Bernard Stiegler', *Études digitales* 9/1 (2020), *Capitalocène et plateformes. Hommage à Bernard Stiegler*, 329–52, 344–5, available at: https://classiques-garnier.com/etudes-digitales-2020-1-n-9-capitalocene-et-plateformes-hommage-a-bernard-stiegler-l-idiodiversite-le-sens-et-la-raison-de-traduire.html.
22 Video by Gaétan Robillard, available at: https://www.ingenieur-imac.fr/realisations/penser-improviser; website realized by Simon Lincelles, available at: http://penserimproviser.org.
23 ImproteK system, introduced by Marc Chemillier, musicologist at the École des Hautes Études en Sciences Sociales.
24 In computer music, musicians are used to build up 'delay' (cf. the MSP Delay Tutorial 1: Delay Lines, available at: https://docs.cycling74.com/max7/tutorials/15_delaychapter01).
25 Bernard Stiegler & Collectif Internation, *Bifurquer: 'Il n'y a pas d'alternative'*, précédé d'une lettre de Jean-Marie Gustave Le Clézio, suivi d'une postface de Alain Supiot, Paris: Les liens qui libèrent, 2020, in particular 'Chap. 4: Recherche contributive et sculpture de soi', 157–77 (Bernard Stiegler & The Internation Collective (ed.), *Bifurcate: 'There Is No Alternative'*, preceded by a letter from Jean-Marie Le Clézio, with an afterword by Alain Supiot, and a lexicon by Anne Alombert and Michal Krzykawski, trans. Daniel Ross, London: Open Humanities Press, 2021, available at: http://www.openhumanitiespress.org/books/titles/bifurcate/, '4. Contributory Research and Social Sculpture of the Self', 119–33; see also Anne Alombert & Michał Krzykawski, 'Vocabulaire de l'Internation', as well as Maël Montévil's contribution to this collection, 'Plaine Commune, contributive learning territory'.

26 Gregory Bateson, 'The cybernetics of "self": A theory of alcoholism', *Psychiatry: Journal for the Study of Interpersonal Processes* 34/1 (1971): 1–18.
27 Winnicott, *Jeu et réalité: L'espace potentiel*, trad. de l'anglais par Claude Monod et J.-B. Pontalis, Préface de J.-B. Pontalis, Paris: Gallimard, 1975, passim.
28 Robert Spaemann, *Bonheur et bienveillance: Essai sur l'éthique*, trad. de l'allemand par Stéphane Robilliard, Paris: Presses Universitaires de France, 1997.
29 Emmanuel Belin, *Sociologie des espaces potentiels: Logique dispositive et experience ordinaire*, 180, Bruxelles: De Boeck, 2002. The concept of 'benevolent setting' (benevolence in an organological design) is related to Winnicott's 'transitional object'. A setting qualifies as benevolent if it is open to desire and libidinal investment.
30 Emmanuel Belin, 'De la bienveillance dispositive (Extrait de sa thèse de sociologie, choisi et présenté par Philippe Charlier et Hugues Peeters)', *Hermès: La Revue* 25/3 (1999): 243–59, 256.
31 Belin, 'De la bienveillance dispositive', 257.
32 Gilbert Simondon, *L'individuation psychique et collective à la lumière des notions de forme, information, potentiel et métastabilité*, préface de Bernard Stiegler, Paris: Aubier, 2007 (1989), 115.
33 Emotional computing is the study and development of systems and devices that can recognize, interpret, process, and simulate human affects. Cf. 'Affective computing', *Wikipedia*, available at: https://en.wikipedia.org/wiki/Affective_computing.
34 Simondon, *L'individuation psychique et collective*, 119.
35 On the so-called hermeneutic and negentropic web, see Bernard Stiegler (dir.), *La Toile que nous voulons*, Limoges: FYP éditions, 2017.
36 See the short text 'SocialWeb' published on the IRI website (12 March 2012), available at: https://www.iri.centrepompidou.fr/projets/socialweb/. Cf. Gilles Deleuze, 'Post-scriptum sur les sociétés de contrôle', *L'autre journal* 1 (May 1990); republished in Gilles Deleuze, *Pourparlers: 1972–1990*, Paris: Les Éditions de Minuit, 1990, 240–7.
37 Individuation proceeds in Simondon from a transductive relation – between a pre-individual potential and its associated milieu – to a non-ontological interpretation of the *Umwelt* in Jakob J. von Uexküll.
38 Georges Canguilhem, *Le normal et le pathologique*, Paris: Presses Universitaires de France, 1966.
39 Clément Morlat, Théo Sentis, Olivier Landau, Anne Kunvari, & Vincent Puig, 'Économie de la contribution et gestion des biens communs', *Imaginaire Communs* (1 March 2021), available at: https://anis-catalyst.org/imaginaire-communs/economie-de-la-contribution-et-gestion-des-biens-communs/.
40 Antonella Corsani, *Chemins de la liberté: Le travail entre hétéronomie et autonomie*, Vulaines sur Seine, Éditions du Croquant, 2020.
41 Antoinette Rouvroy & Bernard Stiegler, 'Le régime de vérité numérique: de la gouvernemantalité algorithmique à un nouvel État de droit', *Socio* 4 (2015): 113–40, available at: http://journals.openedition.org/socio/1251.
42 Alexandre Rambaud, 'Le modèle comptable CARE/TDL: une brève introduction', *Revue française de comptabilité* 483 (2015); the abstract is available at: https://hal.archives-ouvertes.fr/hal-01253482.

Chapter 13

PLAINE COMMUNE, CONTRIBUTIVE LEARNING TERRITORY

Maël Montévil

Plaine Commune is the location of an experiment that is central to the book *Bifurquer*.[1] The tryout builds on earlier philosophical work by Stiegler and Ars Industrialis, the association he co-founded. In this contribution, I will go over some elements of the Plaine Commune programme.

Plaine Commune is a district of nine towns in the north of the greater Paris region, in the Seine-Saint-Denis department. This Parisian suburb has several striking features. The Basilique Saint-Denis is a prominent Christian place of worship from the mid-fifth century onwards, and contains the tombs of many French queens and kings. In the mid-nineteenth century, this suburb was heavily industrialized, and became part of the Parisian red belt, with a solid communist influence. Today, it is a mostly de-industrialized region and a low-income part of Grand Paris. It is also a landing point for many immigrants, notably from former French colonies. Some areas in the region managed to attract the implantation of company headquarters and large infrastructures like the Stade de France (the largest stadium in the Paris metropolitan area). However, these development projects have difficulty meeting the needs of the inhabitants, and, accordingly, they are concerned by the planned infrastructures of the Paris Olympic game. Plaine Commune, and Seine-Saint-Denis generally have a young and creative population that suffers from unemployment, poverty and urban landlock.

The president of the Plaine Commune local authority, Patrick Braouezec, had been following the work of Stiegler and Ars Industrialis for several years, when he asked them to launch an experiment in Plaine Commune. The main aim was to work out and test a contributive income scheme, in connection with a method of contributory research. A call for research projects was subsequently launched in 2017. Put briefly, the contributive income scheme installs a regime whereby people are employed for part of the year and are paid to develop their knowledge for another part of the year. Some such arrangement already existed in France, but was only applicable to live artists, the so-called *régime des intermittents du spectacle* (regime for intermittent entertainment workers). Contributive income might frame any kind of activity; however, the time outside employment requires more specific setups than the pre-existing *régime*, notably the relationship with academic research.

The reasons for elaborating this scheme are manifold. Firstly, automation brings about a tendential fall in employment rates. Even putting aside the ensuing social disasters, a decline in employment leads to a contradiction in consumer capitalism since the system's solvency requires mass consumption – and debt can only go so far to ease a growing surplus. Secondly, automation has intrinsic limitations. Stiegler considers automation to be powerful and in a sense necessary, like in the automatisms of an actor playing his role in a live performance to the automation of a production chain or even to biology. However, automatism also leads to stupidity (*bêtise*), and a rational use of automation should include the means to de-automate, that is to say, the ability to opt out of automaticity whenever it goes wrong. The computational optimizations in the current economic and industrial paradigm tend to neglect this dimension of human activity, with all the dire consequences this neglect entails. Thirdly, Stiegler and Ars Industrialis investigated the role played by the Roman *otium*, a time that is free from productive constraints, by contrast with the *negotium*, the time of productive employment – and let us note that Ars Industrialis' unofficial, initial name was *otium*. Stiegler has often emphasized the significance of the figure of the amateur in all domains, from the arts to technology.[2] The conclusion that imposes itself is that, in a way, work is mostly performed outside employment: as current management focuses single-mindedly on optimization criteria,[3] employment tends to destroy work in collapsing it on the synchronic dimension of human activity, at the expense of its diachronic dimension.

A central reference for the contributive learning programme is the work of the Indian economist Amartya Sen. He observed how the life expectancy in residents of Harlem, New York, had dropped below the expectation of the famine-struck Bangladesh population, at least as men are concerned – death in childbirth was still seriously reducing women's longevity. Amartya Sen explained this remarkable difference in terms of the concept of capacity: practical knowledge and the material situation enabling inhabitants to exert them. Stiegler built on this concept in conjunction with the concept of proletarianization in Karl Marx's work. Proletarianization is the loss of knowledge that tends to result from the transfer of knowledge into a technological device, to begin with the assembly lines that absorbed craftsmanship, as Marx witnessed in his days. Today, proletarianization is not limited to production. Stiegler argued that it has entered everyday life via consumer capitalism, and social life via smartphones and social networks of many kinds. Following the development of digital technologies, proletarianization takes place at an unprecedented speed. This accelerated technological development is most disruptive as it creates a situation in which societies are no longer able to own and appropriate technologies through the elaboration of new knowledge, sciences, regulations and law.[4] This situation is utterly untenable while it leads to an increase in entropies, not only in the domains of physics and biology, but also in human society since it should be reckoned that knowledge is precisely what may delay the increase of entropy, somewhat in the way biological organizations and their normativity do in living beings.[5]

In this context, the overarching aim of the contributive economy, specifically of the Plaine Commune programme, is to develop a dynamic of capacitation, whereby new technologies can be adopted critically and transformed when needed. As may be surmised from the above, this goal entails the overcoming of the effects of both the proletarianization processes that have been going on for some time, and today's disruption. To this end, it cannot be adequate to lean on spontaneous self-organization, both because of the loss of knowledge already suffered as proletarianization increased, and due to the pace of technological change. Capacitation necessitates, rather, the invention of a new relationship between academics, professionals and inhabitants. The goal should be to create groups in which inhabitants feed their experience and academics bring formalized knowledge, in the awareness that such knowledge falls short in grasping and addressing the plights encountered locally – but certainly can contribute to this aim. In this way, the group can operate as a research collective that aims to develop both practical and theoretical knowledge with regard to the situation at hand. The way different actors of the group position themselves will of course differ; this does not alter the fact that in facing disruption, they all take a research stance so as to enter into the specific questions that arise.

In Plaine Commune, several questions are investigated with this research method.

1. The contributive clinic looks into young children parenting in the novel situation where digital media are everywhere, including in the hands of babies and toddlers. The question underlying the work done at the health centre concerns epiphylogenesis and its disruption. Its modus operandi is detailed below.
2. Digital Urbanity in game-mode (*Urbanité numérique en jeux*) aims to develop a new practice of digital technologies so as to address urban changes and to facilitate the inhabitants' adoption of urbanity in the context of the forthcoming Olympic Games (2024). The work group makes use of Minetest – a free software game similar to Minecraft – as an interface to exercise urban modelling in middle and high schools. This programme is developed in partnership with Éducation Nationale (Académie de Créteil), urbanists (O'zone) and sport federations and associations, among others. Inhabitants try to take ownership of their urban milieu and instigate simultaneously a culture of working with digital modelling around concrete issues. This project takes place in the context of the forthcoming Olympic Games in Paris (most infrastructures are not built in the city of Paris but in the Seine-Saint-Denis suburb), where the future urban role of these infrastructure works remains a concern. In this project, the work articulates several groups:
    a) the core, transdisciplinary group (including teachers, informaticians, urbanists, architects, designers),
    b) teachers committed to implementing this project in their school and who underwent specific training as part of the project,
    c) and, of course, schoolchildren participating in the project.

The philosophical question that underlies this initiative concerns the question of urbanity. To that end, Stiegler reworked the concept of 'the right to the city' (*droit à la ville*) of the Marxist philosopher Henri Lefebvre, a concept that has traction in local politics.

3. The work group in Contributive economy explores the set-up of an institutional framework to organize a contributive income policy and to assess its economic benefits. A critical underlying question for this programme is how to integrate the calculable and the incalculable in investment decisions that infringe the ideological premise that only calculable processes count in matters of finance and management (even in academia, based on bibliometrics). To date, this group has primarily carried out theoretical work through the participation of academics and professionals. It notably organizes a seminar in the Caisse des Dépots, the French public investment bank.

Several other projects are in incubation. The emergence of projects is a complex and lengthy process since it requires both an academic interest in the specialities required, the involvement of public or private institutions, the interest of inhabitants and adequate funding. One such emerging project envisions to work on gastronomy sensu Brillat-Savarin, that is, on everything concerning humans insofar as they feed themselves, from recipes to geopolitics and public health.[6] A central underlying question concerns taste, specifically with reference to Nietzsche. Another project targets the conversion of combustion engine cars into electric cars by building on the advanced practical knowledge that local 'street mechanics' possess.

Stiegler's perspective on these projects was not just to consider them one by one, since he would then reconfirm the division of labour – as a particular kind of functional isolation sensu Shaj Mohan and Divya Dwivedi[7] – that has led to the organization of artificial stupidity and proletarianization. He attempted, rather, to reticulate the many projects, notably under the umbrella of the question of generations and of the transgenerational.

To conclude this paper, I will enter into some aspects of one of these projects that focuses directly on the transgenerational, the contributive clinic. This programme builds on previous work by Stiegler on epiphylogenesis and, particularly, on attention. In the Plaine Commune programme, it was a critical question, and Stiegler started to address and develop it, in collaboration with Anne Alombert, into a new approach of psychotherapy in the context of disruption. After reading an earlier book,[8] Stiegler was contacted by Marie-Claude Bossière, a child psychiatrist (*pédopsychiatre*, in the singular French tradition that emerged after the Second World War) who is part of a group of clinicians who alert on the consequences of young children's overuse of digital media and, in general, are concerned by the question of parenting, once smartphones and tablets enter everyday life.[9] As I was joining IRI at the time, I would bring in the question of entropy from the perspective of biological organizations and their disruptions, notably through the concept of anti-entropy to accommodate organization from both a systemic and diachronic perspective.[10] This approach enabled us to draw comparisons between the use of digital media, endocrine disruptors or the disruption of plants-pollinators networks at the level of ecosystems. Anne Alombert developed the project's methodological and philosophical aspects that are connected to the role of technical supports in psychological or noetic activities.[11]

After discussion with several actors of the Plaine Commune territory, and after exploring several institutional possibilities, the group opted to cooperate with an existing Protection Maternelle et infantile (PMI; English: Center for the protection of mothers and young children). PMIs are public medical centres that monitor children from the age of 0 to 6, with a focus on prevention and parental advice. Late in 2018, the public services of Saint-Denis spread our call to cooperate, and the PMI Pierre Semard, located in one of the poorest parts of Saint-Denis, accepted it. We organized work in successive steps. The first step only involved the professionals of the PMI; next, parents who were former patients of Marie-Claude Bossière and had overcome screen overuse problems, joined the group; finally, the group opened its doors to parents from the area.

Considering the rapidity of the change smartphone and tablet technologies are subject to, Stiegler took the stance, in his relationships with professionals and parents, that we are all lost as it comes to taking care of kids, and, more generally, as our everyday and working life are concerned. In this regard, academics are not just bringing in scientific knowledge, they contribute in drawing from their own experience and the difficulties they encounter. Parents and professionals, for their part, can build on academic knowledge so as to understand better the challenges they are facing, and contribute in turn to ongoing academic research. Since they are having similar problems and are lacking the requisite knowledge to tackle them, all group actors are primarily occupying symmetric positions that are complemented with asymmetries due to their having different knowledge, experience and duties. For example, and to put it somewhat bluntly, Stiegler had obviously a lot to bring in, as for concepts in general and the relationship between technics and *noēsis* in particular; yet, he recognized the authority of Marie-Claude Bossière as for the therapeutic devices, and to the authority of professionals and parents as for the local situation.

The work was organized through the articulation of two groups: the PMI group, and the group of participants in a monthly academic seminar at IRI – most of them academics and external psychologists. In view of the integration of academic research and the research carried out by the PMI group, the seminar was followed by a transfer moment during which its course and content could be shared by the PMI group. The seminar built on texts that might contribute to a better understanding of the ways in which the development of young children gets disrupted and therapeutic devices may be co-constructed. This arrangement enabled us from the very start to get into challenging texts, such as Bateson's essay on alcoholism from a cybernetic perspective, which the PMI group did actually read in a systematic way.[12] During these meetings, Stiegler contributed chiefly as a philosopher, introducing concepts and working on them by relying on text material. Stiegler's summons to parents and professionals was, in short, to try and contribute as researchers, and the group did heed his call. The configuration of the seminar has changed in the context of the Covid-19 pandemic. PMI professionals wished to follow the academic seminar directly and the shift to videoconferencing provided an opportunity to overcome practical difficulties to this end.

The integration of parents to the PMI group and the work it carried out, made use of a specific device: relatively short videos serving as a basis for discussion. The videos have been uploaded to the website of the work group.[13] After watching the videos, the discussions combined the insights of practitioners and parents with the scientific and philosophical input. The use of videos in the therapeutic device exemplifies the pharmacological perspective on technologies: screens and digital media are both remedies and poisons, and there is, therefore, no contradiction in making use of screens to address issues caused by screens. The PMI professionals made the notion of *pharmakon* their own and wielded it in their preventive work and discussions with parents who were not participating in the group. This perspective on technology was of value during the Covid-19 confinement periods, when work had to be done by videoconference. Indeed, during the first lockdown, rather than freezing all activity, the group chose to double the frequency of its meetings, from once to twice a week. During this period, the paradigm of institutional psychotherapy, in which patients take care of the institution, was critical.

In the context of Covid-19, and following the death of Stiegler, the group has obviously faced hardships. It still holds strong, though: it is currently working towards the transfer of its knowledge to other professionals and parents in Saint-Denis and surrounding towns. It was one of Stiegler's intentions to recreate a *philia* in the industrialized economy and this has worked out particularly well in the contributory clinic's case.

Notes

1. Bernard Stiegler & Collectif Internation, *Bifurquer: 'Il n'y a pas d'alternative'*, précédé d'une lettre de Jean-Marie Gustave Le Clézio, suivi d'une postface de Alain Supiot, Paris: Les liens qui libèrent, 2020; Bernard Stiegler & The Internation Collective (ed.), *Bifurcate: 'There Is No Alternative'*, preceded by a letter from Jean-Marie Le Clézio, with an afterword by Alain Supiot, and a lexicon by Anne Alombert and Michal Krzykawski, trans. Daniel Ross, London: Open Humanities Press, 2021, available at: http://www.openhumanitiespress.org/books/titles/bifurcate/.
2. Bernard Stiegler, 'Le temps de l'amatorat. Entretien avec Éric Foucault', *Alliage: Amateurs?* 69 (2011): 161–79.
3. Bernard Stiegler, 'Work as the Struggle against Entropy in the Anthropocene', trans. Daniel Ross, *Harvard Design Magazine: No Sweat* 46 (2018): 177–81, available at: http://www.harvarddesignmagazine.org/issues/46; https://www.academia.edu/38945541/Bernard_Stiegler_Work_as_the_Struggle_against_Entropy_in_the_Anthropocene_2018_.
4. Bernard Stiegler, *The Age of Disruption: Technology and Madness in Computational Capitalism*, followed by *A Conversation about Christianity with Alain Jugnon, Jean-Luc Nancy and Bernard Stiegler*, trans. Daniel Ross, Cambridge: Polity, 2019.

5   Cf. Bernard Stiegler, *The Neganthropocene*, ed. and trans., and with an introduction by Daniel Ross, London: Open Humanities Press, 2018; Maël Montévil, 'Sciences et entropocène. Autour de *Qu'appelle-t-on panser?* de Bernard Stiegler', *EcoRev'* 50/1 (2021): 109–25, eprint available at: https://montevil.org/assets/pdf/2021-Montevil-Stiegler-Sciences-Entropocene.pdf, https://montevil.org/publications/articles/2021-Montevil-Stiegler-Sciences-Entropocene/; DOI available at: 10.3917/ecorev.050.0109.
6   Jean Anthelme Brillat-Savarin, *The Physiology of Taste: Or Meditations on Transcendental Gastronomy*, trans. and ed. M.F.K. Fisher, introduction by Bill Buford, New York: Vintage Books, 2009.
7   Divya Dwivedi. 'Through the Great Isolation: *Sans-colonial*', *Philosophy World Democracy* (22 November 2020), available at: https://www.philosophy-world-democracy.org/through-the-great-isolation.
8   Bernard Stiegler, *Taking Care of Youth and the Generations*, trans. Stephen Barker, Stanford: Stanford University Press, 2010.
9   Daniel Marcelli, Marie-Claude Bossière & Anne-Lise Ducanda, 'Plaidoyer pour un nouveau syndrome « Exposition précoce et excessive aux écrans » (EPEE)'. In: *Enfances & Psy* 79/3 (2018): 142–60, available at: https://www.cairn.info/revue-enfances-et-psy-2018-3-page-142.htm; DOI available at: 10.3917/ep.079.0142.
10 Maël Montévil, 'Entropies and the Anthropocene crisis', *AI and Society* (May 2021), available at: https://montevil.org/publications/articles/2021-Montevil-Entropies-Anthropocene/; DOI available at: 10.1007/s00146-021-01221-0.
11 Anne Alombert, 'Faire du choc une chance?', *Zone Critique* (1 May 2020), available at: https://zone-critique.com/2020/05/01/faire-choc-chance/.
12 Gregory Bateson, 'The cybernetics of "self": A theory of alcoholism', *Psychiatry* 34/1 (1971): 1–18.
13 The website of the *Atelier Contributif – Écrans* on screens and young children is available at: https://atelierecrans.wordpress.com/.

Chapter 14

TOWARDS A BIFURCATION: INTERNATION AND INTERSCIENCE IN THE TWENTY-FIRST CENTURY

Anne Alombert

Introduction

The notion of internation was initially coined by Marcel Mauss, in a book written in 1920 called *The Nation*, in which the anthropologist studied the history of nations and raised the question of international relationships. The notion of internation meant to go beyond the alternative between a nationalism that isolates nations, and an internationalism that ignores national specificities.[1] According to Mauss, writing at the time of the institution of the League of Nations, two years after the end of the First World War, the ideal of internationalism should not lead to a 'supranation' that absorbs all nations, or to a 'cosmopolitism' that denies the nation as a social fact, but should constitute the pillar of an internationalism articulating national diversities and uniting nations instead of erasing them.[2]

While Mauss was developing his reflexions on the internation, Albert Einstein was insisting on the necessity to create an international organization of science, which would allow the scientists and 'intellectual workers' of the world to work together for the benefit of international interests.[3] At that moment Einstein was involved in the International Commission of Intellectual Cooperation. This panel was created in 1922 into the League of Nations and was then chaired by Henri Bergson. According to Einstein, its main goal was to bring together the intellectuals, artists and scientists of all nations, nationalist tendencies notwithstanding.[4]

Reflecting on international relations and on academic cooperation between the two World Wars, Mauss, Einstein as well as Bergson were outlining and setting up an international organization for peace, in a context in which the mortal character of civilizations[5] and the ambiguity of technoscientific progress had become obvious to everybody. One century later, at the beginning of the twenty-first century, Bernard Stiegler reactivates these questions, insisting on the necessity of both an 'internation' and an 'interscience' in the context of what he describes as a 'global economic war'[6] which, according to him, destroys all forms of local knowledge (know-hows, know how to live and theoretical knowledge) and inevitably leads to the 'return of nationalisms'[7] reacting to the 'functional sovereignty'[8] that is imposed by global digital platforms.

Internation in the twenty-first century: An alternative path beyond authoritarian nationalisms and the 'dunctional sovereignty' of digital platforms

More than twenty years ago, in a text about globalization and 'teletechnologies', Jacques Derrida already explained that the demand for frontiers, identities and roots could be read as a defensive reaction trying to compensate for the dislocations, expropriations and uprootings provoked by the planetary extension of new information and communication technologies and new technoscientific developments:

> A general hypothesis is therefore insufficient. Still, insufficient as it remains, I believe it is necessary in that it appeals to the technological process in so far as this process (although not only) takes the general form of expropriation, dislocation, deterritorialization. [...] Today, we are witnessing such a radical expropriation, deterritorialization, delocalization, dissociation of the political and the local, of the national, of the nation-state and the local, that the response, or rather the reaction, becomes: 'I want to be at home, I want finally to be at home, with my own, close to my friends and family'. [...] The more powerful and violent the technological expropriation, the delocalization, the more powerful, naturally, the recourse to the at-home, the return towards home.[9]

Nowadays, these 'teletechnological' dislocations, expropriations and uprootings have been concretized through what Frank Pasquale calls the 'functional sovereignty' of digital platforms and giant tech companies, which bypass territorial economic regulations and short-circuit local political systems through 'smart' innovations:

> [Digital firms] are no longer market participants. Rather, in their fields, they are market makers, able to exert regulatory control over the terms on which others can sell goods and services. Moreover, they aspire to displace more government roles over time, replacing the logic of territorial sovereignty with functional sovereignty. In functional arenas from room-letting to transportation to commerce, persons will be increasingly subject to corporate, rather than democratic, control.[10]

Through their promises of economic growth and technological progress, giant tech companies settle a competition between territories and convince territorial administrations to transfer their power to them. They disrupt local economic activities and turn territories into experimental grounds for their innovations, in order to maximize their techno-economical performances independently of any democratic mandate. According to Pasquale:

> Solutions to Amazon's power will, no doubt, be hard to advance as a political matter – consumers like 2-day deliveries. But understanding the bigger picture here is a first step. [...] And only political organization can stop its functional sovereignties from further undermining the territorial governance at the heart of democracy.[11]

According to Bernard Stiegler, this functional sovereignty, which implies an automation and a smartification of everyday lives and environments, not only bypasses the 'political sovereignty of territories' but also destroys local 'singularities' through the standardization of ways of life. Automated systems thus eliminate the local forms of knowledge (know-hows, know how to live, theoretical knowledge)[12] which give their specificities and their vitality to the localities. What Stiegler sees threatened is the singular diversity of the spheres of knowledge that make up noetic life, the 'noodiversity'[13] of which tends to be progressively eliminated through the 'inextricably economic and technological war'[14] that we call globalization. In this context, the function of the internation is to open up an alternative path to both globalization and nationalist reactions, through the protection and cultivation of local spheres of knowledge and through the reconstitution of solvent and diversified localities:

> Noodiversity is the starting point of the internation: it is only through the *cultivation of the diversity of knowledge*, all forms of which are universalizable, but which, on the other hand, can be cultivated only locally, that it is possible to limit the anthropy produced by exosomatic evolution. It is precisely for this reason that the internation, as a higher complex exorganism of reference, must *protect* this diversity of local knowledge against any attempt to homogenize it, that is, against all entropic and anthropic tendencies.[15]

The internation thus designates an agreement, a consensus and a network between various open localities (which can be nations, regions, metropolises, territories, etc.), united in the common concern to design, and to experiment with economic models that promote local knowledge and arts of living, and thus take care of the biosphere.

Interscience in the twenty-first century: The reconstitution of public power and public space through contributory research and contributory digital technologies

Such experimentation calls for 'contributory research',[16] through which academic searchers come to work with local inhabitants (citizens, associations, political representatives, economic or industrial actors, professionals, etc.) so as to elaborate new spheres of knowledge and set up new social organizations. These knowledges and organizations should favour the adoption of the technical transformations and industrial innovations that are today primarily socialized through a marketing effort that destroys 'psychic apparatuses' and 'social systems'.[17] This explains why the internation is also an 'interscience'[18] entailing the conjoined individuation of different academic disciplines, and of these disciplines and the 'outside' of the university in particular, through which theoretical questions come to relate to concrete, local problems. This 'interscientific' approach works towards a rearticulation of different kinds of knowledge (theoretical knowledge, practical knowledge, political knowledge, professional knowledge) through the establishment of partnerships between universities and civil society (in both its economic and political dimension):

> the partnership between the university and economic civil society must also pass, contractually and necessarily, through a *partnership of the university with political civil society*. 'Politics' here means: associations of citizens, amateurs, activists and residents who are encouraged to work together with the academic world.[19]

The main objective of such processes of contributory research is to establish communities of amateurs or communities of knowledge on the basis of which people may capacitate each other while sharing different kinds of knowledge. In this way they may invent new ways of doing, living and thinking with digital technologies:

> it is essential that the university invent a new relation to its outside (and, through that, to the question of its milieu, and not just of the 'environment', whether physical, economic, political or mental), via the theory and practice of *hypomnēmata* and in relation to the community of amateurs, that is, through the development of contributory research.[20]

The 'academic internation' is hence defined as a process of co-individuation between many universities from all over the world. While participating in the conduction of contributory research at local levels, these institutions work together at an international level in order to exchange and discuss their ideas and experiments through a process of collective individuation:

> The *academic internation* could and should, therefore, become the catalyst for a new process of collective individuation at the global level, enabling negotiations to be conducted towards a global civil peace treaty.[21]

Based on scientific work and social experiment, these negotiations could reconstitute collective protentions at a planetary scale. These may benefit the replacement of conflicts of war by rational controversies and public deliberations on the future of biodiversity and noodiversity, as these are threatened by the

Anthropocene and digital disruption. Deliberation and controversy require, however, the establishment of a new public space. Based on the reinvention of the 'digital system of publication',[22] such a shared space should favour processes of contributory annotations, indexations, categorizations as well as collective debate and discussions. Contributory research implies therefore 'the design and development of systems of contributory editorial production'[23] in view of the arrangement, at an international level, of a political and public space that effectively counterbalances the hegemony of giant tech platforms. Contributory digital tools should allow for the reticulation of diverse localities through the circulation of different spheres of knowledge. These 'savoirs' are invariably localized yet likely to de-territorialize as they are exchanged, conflict and enrich each other in the course of a transindividuation-process constituting a new public power at a planetary scale.

Bernard Stiegler's last project: An attempt to bring about an internation and interscience

It is in this context that the Internation collective was initiated by Bernard Stiegler in 2018, six years after the writing of *States of Shock*, during a symposium he organized with Hans Ulrich Obrist in London.[24] This collective was composed of academic searchers in many different fields (philosophy, biology, mathematics, economy, law, geography and sociology) and of professionals in different domains, including doctors, artists, engineers, designers, business men and women. The aim of its contributors was to work together for two years, in small workshops and during punctually set symposiums, so as to provide a shared comprehensive scientific analysis of the current global situation that is usually called the Anthropocene, and to formulate proposals in order to initiate a bifurcation to go beyond this situation.[25]

To sum up the content of this transdisciplinary work, we might say that the reflections of the Internation collective proceed from Bernard Stiegler's thesis that the Anthropocene can be described as an 'Entropocene'[26] and can hence be explained in terms of an augmentation of entropy rates in different fields:

1. in the thermodynamic field through the exploitation of natural resources;
2. in the biological field through the elimination of biodiversity;
3. in the psycho-social field through the destruction of knowledge and the standardization of ways of living.

It follows from this thesis that if we want to go beyond this 'entropocenic' situation, we must set up, and experiment with a new economic model that is to be explained in terms of both (1) the fight against entropy-production and (2) the valorization of anti-anthropic activities, defined as work activities during which people practice all kinds of knowledge – theoretical knowledge, technical knowledge (know how), practical or existential knowledge (know how to live) – it being understood that such practices invariably take place at local levels. The different groups of the Internation collective all set out from this broad argument in an attempt to project and develop it in different domains: of epistemology, education, economy, design, ethics, international politics, urbanity, etc.

The results of this transdisciplinary work were publicly presented in Geneva on 10 January 2020 at the occasion of the centennial of the League of Nations. The purpose of this event was to launch a discussion at the level of the United Nations on the concerns addressed by the Internation collective, so as to convince the United Nations to set up an international programme of contributory research on the questions raised by the Entropocene-thesis. Such research could elaborate the issues at a theoretical level and articulate them at the same time with concrete social experimentation, in different 'laboratory territories' or 'learning territories'. The proposals of the Internation collective were first published online,[27] and subsequently turned into a book entitled *Bifurquer*,[28] published in France in June 2020. Since then both the Italian[29] and English translations have appeared, while the Polish translation is forthcoming.[30] Thanks to these multiple translations, we can hope that the analyses and propositions offered in this book will be interpreted, appropriated, discussed and transformed in diverse localities, and will thus open a path towards an internation to come. As a tertiary retention or a hypomnesic support carrying the memory of past experiences that await reactivation, re-individuation and renewal, we may hope that this collective book will work beyond itself, 'bear[ing] within it that which does not resemble it, but which it alone has made possible'.[31]

Notes

1   'An internationalism worthy of the name [...] is not a denial of the nation. It situates it. The internation would be the opposite of such a-nationism. It is also, therefore, the opposite of nationalism, which isolates the nation', Marcel Mauss, *Œuvres*, Paris: Minuit, 1969, 630, quoted in Bernard Stiegler, *States of Shock: Stupidity and Knowledge in the 21st Century*, trans. Daniel Ross, Cambridge: Polity Press, 2015, 179. See also Bernard Stiegler & The Internation Collective (ed.), *Bifurcate: 'There Is No Alternative'*, preceded by a letter from Jean-Marie Le Clézio, with an afterword by Alain Supiot, and a lexicon by Anne Alombert and Michal Krzykawski, trans. Daniel Ross, London: Open Humanities Press, 2021, available at: http://www.openhumanitiespress.org/books/titles/bifurcate/, par. '67. Nations, globalization, internationalism and the internation', 136–40.

2   'Cette composition de rapports définit ce qu'il faut alors appeler l'internationalisme, qui est à l'opposé du nationalisme, puisqu'il récuse par principe le fait que la nation puisse être conçue comme un isolat. Mais il est aussi à l'opposé du cosmopolitisme, qui table sur une disparition du fait national [...]', Bruno Karsenti, 'Une autre approche de la nation: Marcel Mauss', *Revue du MAUSS* 36/2 (2010): 283–94.

3   Albert Einstein, *Ideas and Opinions*, trans. Sonja Bargmann, New York and Avenel: Wing Books, 1954, 83–5.

4   'This commission was to be a strictly international and entirely non-political body, whose business it was to put the intellectuals of all the nations, who were isolated by the War, in touch with each other. It proved a difficult; for it has, alas, to be admitted that [...] the artists and men of learning permit themselves to be governed by narrow nationalism to a far greater extent than the men of affairs.', Einstein, *Ideas and Opinions*, 83–5.
5   'We later civilizations [...] we too know that we are mortal', Paul Valéry, *Crisis of the Mind* (1919), trans. Denise Folliot & Jackson Mathews, available at: https://en.wikisource.org/wiki/Crisis_of_the_Mind.
6   On the questions of 'internation' and 'interscience' in Stiegler's thought, see Stiegler, *States of Shock*, 'Chap. 8: Internation and Interscience', 173–202, and 'Chap. 9: Interscience, Intergeneration and University Autonomy', 203–20.
7   Stiegler & Internation, *Bifurcate*, 136 and 161.
8   Frank Pasquale, 'From Territorial to Functional Sovereignty: The Case of Amazon', *Law and Political Economy* (6 December 2017), available at: https://www.opendemocracy.net/en/digitaliberties/from-territorial-to-functional-sovereignty-case-of-amazon/.
9   Jacques Derrida & Bernard Stiegler, *Echographies of Television: Filmed Interviews*, trans. Jennifer Bajorek, Cambridge: Polity Press, 2002, 79–80.
10 Pasquale, 'From Territorial to Functional Sovereignty: The Case of Amazon'.
11 Pasquale, 'From Territorial to Functional Sovereignty: The Case of Amazon'.
12 Cf. Stiegler & Internation, *Bifurcate*, par. '31. Towards collective territorial agreements and new forms of urban intelligence: challenges of contributory urbanity', 75–7.
13 Cf. Stiegler & Internation, *Bifurcate*, par. '76. The effects of proletarianization and the future of public authority in the Anthropocene', 151–3.
14 Stiegler & Internation, *Bifurcate*, 152.
15 Stiegler & Internation, *Bifurcate*, 140.
16 On contributory research, see Stiegler, *States of Shock*, par. '77. Contributory research beyond the inside and the outside of the university', 203–6 and Stiegler & Internation, *Bifurcate*, Ch. '4. Contributory Research and Social Sculpture of the Self', 119–33.
17 'Defined as "creative destruction," industrial innovation leads inevitably to the short-circuiting of the adoption of technical individuation by psychosocial individuation, generating massive, systemic and ruinous disindividuation of psychic apparatuses and social systems', Stiegler, *States of Shock*, 188.
18 Stiegler, *States of Shock*, par. '68. Internation and interscience', 185–8.
19 Stiegler, *States of Shock*, 207.
20 Stiegler, *States of Shock*, 204.
21 Stiegler, *States of Shock*, 180.
22 Stiegler, *States of Shock*, 181.
23 Stiegler, *States of Shock*, 213. On contributory digital technologies and contributory editorial systems, see also Stiegler, *States of Shock*, par. '80. The gay science – metadata and metalanguage' and '81. The editorial and publishing industries of scholarly and scientific society', 209–14 and Stiegler & Internation, *Bifurcate*, Ch. '7 Contributory Design and Deliberative Digital Technologies: Towards Social Generativity in Automatic Societies', 178–94.
24 The recordings of this symposium are available at: https://iri-ressources.org/collections/collection-47.html.
25 The propositions of the Internation collective were progressively published online and are available at: https://internation.world/.

26 'Our main thesis is that the Anthropocene era can be described as an Entropocene era, insofar as it is characterized above all by a process of the massive increase of entropy in all its forms (physical, biological and informational). The question of entropy has, however, been neglected by "mainstream" economics. We therefore believe that a new macroeconomic model, designed to struggle against entropy, is required', Stiegler & Internation, *Bifurcate*, 'Letter from Hans Ulrich Obrist and Bernard Stiegler to António Guterres', 11–13, also available at: https://internation.world/arguments-on-transition/letter-to-guterres/.
27 The propositions of the Internation collective were first published online and are available at: https://internation.world/.
28 Bernard Stiegler & Collectif Internation, *Bifurquer: 'Il n'y a pas d'alternative'*, précédé d'une lettre de Jean-Marie Gustave Le Clézio, suivi d'une postface de Alain Supiot, Paris: Les liens qui libèrent, 2020. For a presentation and summary of the book as well as a glossary of the notions, see A. Alombert et M. Krzykawski, 'Vocabulaire de l'Internation', *Appareil* (3 February 2021), available at: http://journals.openedition.org/appareil/3752.
29 B. Stiegler & Collettivo Internation, *L'assoluta necessità: In risposta ad Antonio Guterres e Greta Thunberg*, traduzione e cura di Sara Baranzoni, Giacomo Gilmozzi, Edoardo Toffoletto e Paolo Vignola, Rome: Meltemi, 2020.
30 B. Stiegler & Internation, *Konieczna bifurkacja: 'Nie ma alternatywy'*, trans. Michal Krzykawski, Milosz Markiewicz, Katowice: Wydawnictwo Uniwersytetu Śląskiego, 2021.
31 Stiegler, *States of Shock,* 264.

Chapter 15

A *SCHOLĒ* FOR THE THUNBERG GENERATION

Victor Chaix

The bleak situation of the Anthropocene for present and future human living conditions fetches unprecedented questions for present and future generations and their relationship with each other. How should younger and older generations living in these turbulent times relate to each other? What responsibility does it ultimately confer to them both, in terms of transmission of knowledge across generations and in terms of their collective re-evaluation of the legacy of the Anthropocene?

The figure of Greta Thunberg, her bold speeches to world leaders and the many members of her generation that had been striking from school with her, around the world, testifies to a deep generational tear in terms of models of aspiration: between what we might call a 'boomer' representation of the course of global history, entailed by the modern idea of 'progress', and the profound disillusion of the 'Thunberg generation' in this system of beliefs.

As Thunberg and climate strikers deplore, a certain dream of eternal growth and technological solutionism to humanity's problems has turned into a climate-nightmare. This realization strikes the Thunberg generation, who becomes increasingly unable to *dream* – in a Stieglerian sense of the word: 'dreams that are realisable, rational dreams'[1] – and very humbly implores its elders to 'speak the truth' and to 'act now' for decent future living conditions in a habitable planet.[2]

The effort has met with little success. The new world or direction that the international youth movement of Fridays For Future and Extinction Rebellion (XR) calls for, is still not here. The significant reverberation they gave to the reports of global scientific institutions – such as recent ones given by the Intergovernmental Panel on Climate Change (IPCC), concerning the acceleration of climate warming, and by the Intergovernmental Science-Policy Platform on Biodiversity and Ecosystem Services (IPBES), concerning the reduction of biodiversity[3] – was not and is still not followed by genuine transformations in the socio-economical dynamics that caused these alarming projections in the first place.

Acknowledging the limits of the efficiency of these movements, Bernard Stiegler and the French literary figure J. M. G Le Clézio launched the Association des Amis de la Génération Thunberg (AAGT; English: Association of the Friends of the Thunberg Generation), together with international scientific personalities and student activists from Fridays For Future and Extinction Rebellion, to work on the conceptual and methodological gaps that hinder – beyond the lack of political willpower – the possibility to generate both locally and systemically the *bifurcations* that are indispensable to meet and face our new reality.

In this paper, I will try to demonstrate how the AAGT is premised on a re-evaluation of both activism and scientific knowledge, that is to say, on the imperative to reconcile sociopolitical action with rigorous thinking and on the urgent necessity to put scientific research and the production of knowledge at the service of viable sociopolitical and economic alternatives.

Despite the tremendous challenges that the newborn association was confronted with during its early months – social-distance measures due to the Covid-19 outbreak, as well as the death of the association's president in August 2020 – these ambitions have started to be put to practice through what we have called the 'School of the Thunberg Generation': a school designed by the philosopher and the most active members of the early association, addressed to the Thunberg generation. Itinerant and online, aiming to settle as diverse intergenerational working groups, this 'school' can be understood in different ways. Here, I will attempt to sum up and analyse the most relevant ideas, experiences and experimentations that have emerged since the officialization of the association on 27 February 2020.

The Thunberg generation & the boomers

The AAGT strives to re-articulate the relationship between generations, torn apart for at least half a century by the media and advertising industry,[4] and more recently, by the collective awareness of the already catastrophic state of the biosphere. In this context, a process of re-articulation is needed between what has been called the 'boomers', often in a pejorative way, and what we call with Stiegler the 'Thunberg generation',[5] who expressed their discontent towards the boomers in mass demonstrations during 2019. We labour to heal this tear, well attested by these very categorizations and this dichotomization of generations.

How come the climate generation differs so much from its boomer counterpart, born closely after the Second World War,[6] that the expression 'OK boomer', coined by the New Zealand parliamentary Chlöe Swarbrick in late 2019, was able to gain tremendous popularity as a credo amongst young climate demonstrators? This question was first posed during a round table at the Entretiens du Nouveau Monde Industriel (ENMI; English: New Industrial World Conference) 2019.[7] The 'Thunberg generation' – as restrictive as this category can be – has been for us a subject of interrogation and contemplation, during the seminars organized by the AAGT in collaboration with the Institute of Research and Innovation (IRI), in the spring and summer of 2020.[8]

During this time, I wrote an article on the AAGT Mediapart blog with the contribution of Stiegler and other friends of the Thunberg generation, in an attempt to 'define' in our own terms this concept of 'Thunberg generation', in all its complexity and variances[9] and in comparison with, for instance, the 'Umbrella generation' of Hong Kong demonstrators or the 'Floyd generation' in the United States, ... as well as with what was increasingly being called, by the French newspaper *Le Monde* amongst others, the 'Covid generation'.[10]

Although a generation is never monolithic and stays inherently full of diverse, sometimes contradictory trends, I attempted to describe the stakes that pertain to the Thunberg generation and what is also inevitably the responsibility of other generations. A first thing that comes to mind when talking about this part of the 'Generation Z' – approximatively born between 1997 and 2010 – is their technological condition: the increasing digitalization of their social interactions and of their perceptions of the world. The 'online generation' is characterized inherently, and first and foremost, by its technological milieu: a philosophical affirmation that Stiegler made across much of his work, leading up to his last *Pharmakon*-seminars.[11]

When it comes to the relation between the 'Thunberg generation' and the boomers, I argued that what is at stake is a conflict between two aspirational models, two distinct representations of the world and their corresponding historical trajectories. The Thunberg generation, which still works on 'modelling' its aspiration, is deeply disillusioned by the sociopolitical model that they inherit, which is centred around the growth of the GDP and the unequivocal faith in a Euro-American centred vision of 'progress', which presently takes the form of techno-solutionism, if not transhumanism.

We can therefore envision the 'Friends' of the Thunberg generation as the specific part of the 'boomer generation' and the post-boomer generation that not only recognize the tragedy that face young and future generations with regard to their future living conditions, but that also have a strong and sincere *will* to support their struggle for a world that could be habitable and desirable. In the AAGT, this will take the form of a transmission and co-elaboration of the theoretical tools and practical schemes that are indispensable to heal societies and set them on viable, and desirable grounds. In this sense, the friends of this generation work and discuss intimately with its members, in a process of getting to know and understand each other. This is what Bernard Stiegler already described in 2008, in his book *Prendre Soin 1. De la jeunesse et des générations* (translated as *Taking Care of Youth and of the Generations*).[12]

Intergenerational transmission and renewal

In many ways, the young association finds its roots in earlier structures and activities. In 2005, Bernard Stiegler launched with three other philosophers – Georges Collins, Marc Crépon and Catherine Perret – and a jurist, Caroline Stiegler, the association Ars Industrialis (AI).[13] This association called for a new kind of industrial politics concerning the 'technologies of the mind and spirit' ensuring that these technologies were not subject to the imperatives of a consumerist economy.

We may interpret Stiegler's decision to merge the new AAGT with the structure of AI during a general assembly of the association on 27 February 2020, as a way to renew what was thought and worked on in AI, by including in this structure young members – which in general have a stronger plasticity of mind and thought – so that the work would be transmitted and its loss prevented. Consciously or unconsciously, Stiegler might have followed Planck's principle, pertaining to the sociology of scientific knowledge:[14] to make his knowledge and theories progress, it is more strategic and wiser to transmit it, renew those through young generations, than to try changing the mind of scientists and their dominant scientific paradigm.

In this sense, the AAGT is more oriented towards this work of transmission between its members than Ars Industrialis, with its numerous mediatized and public events.[15] This reorientation and renewed relation to the media is affirmed in the Association's vocation, when it states the favouring of private work sessions, away from cameras and the media.[16] More recently, in the postface of a book on the subject of education in the twenty-first century,[17] Stiegler seems to justify this 'inward' re-orientation as a way to protect collective thoughts, debates and insightful reflections from the contemporary 'storms of stupidity', which is indeed most acutely observed in the media industry.

Beyond Ars Industrialis and along with the book *Prendre Soin*, Stiegler wrote numerous papers and speeches revolving around the questions of youth and generations, which is a central theme in his rich and complex thoughts on technics. The philosopher also held a series of Pharmakon courses, from 2011 to 2014, particularly addressed at high school students.[18] In this sense, Stiegler's concern for the youth and intergenerational relationships did not emerge ex nihilo, but is rather a constant in his thought. It is just that in the critical context of what he dared to call 'the end of the Anthropocene era'[19] (supposed to give way to the Neganthropocene era), the question of youth and intergenerational solidarity has never been as salient and crucial to philosophy.

Therefore, starting as an association that would defend Greta Thunberg against the persons that want her dead – such as the previous president of the Association des Amis du Palais de Tokyo (art amateurs that support the prestigious museum of the same name, in Paris), which publicly called on Facebook for someone to 'take her down'[20] – it was becoming increasingly clear for Stiegler and the other founding members of the AAGT that a strong intergenerational support was needed, going beyond the figure of Thunberg.

Re-articulation between science and politics, knowledge and action

I would argue that the *raison d'être* of the AAGT is first and foremost to respond to our critical contemporary situation in a way that is worthy of the challenges at stake – in other words, in a rigorously original and ambitiously realistic manner. That is why we argue that science and research, activism and sociopolitical action, should collaborate at once intimately and seriously with each other. The AAGT makes the bet that we should go counter to the idea that complex realities can be made 'easy' through 'communication' techniques. On the contrary, we try to transform what is complex and hardly graspable into a desirable object, thus engaging in a counter-dynamic of what Stiegler called proletarianization – the loss of knowledge, due mainly to emerging technological milieus and consumption capitalism. We work for collective de-proletarianization, not only within the contemporary world of activism, but also in the scientific world itself.

Activists have lost a certain dimension in their actions: the meticulous elaboration of theories and practices that was until recently fundamental in social movements. We attempt in the AAGT to valorize a counter-current activism by *actively* thinking and working for an alternative to what we denounce. As Stiegler called for during the introduction of an Ars Industrialis summer academy in 2011:[21] instead of resolving ourselves to 'contemplate the Anthropocene and its destruction', we should rather attempt to promote what he called a 'quasi-causal engagement in the ignored negentropic possibility'. For this, we should not obey our pulsion but rather follow our desire for truth, even in its politically frustrating aspects.

After our first collaboration in October 2019, Stiegler invited me tacitly to engage in this re-evaluation of activism and political action. For his new book, I had to transcribe the conference 'Knowledge and Politics' he held in 2011 in a Parisian café.[22] In this rather unknown conference, he impelled French left movements to interrogate the status of knowledge in political action. As he deplored, the credit of political action 'collapses' at the beginning of the twenty-first century, due to the lack of founding political work on knowledge – more precisely, on theoretical knowledge, which serves to frame an elaborate critique of our particular sociopolitical context. I found on that occasion a direction to my own activism, which was Stiegler's affirmation that 'an activist thinks and works' and his concomitant injunction to 're-arm thought'.

We thus need to reconsider scientific knowledge and politics as inseparable from each other: scientific knowledge and theories often have considerable political consequences, while political systems or decisions in turn condition – as much as the technological infrastructures and their algorithms in the digital age do – the very production and occultation of certain kinds of knowledge. Therefore, Stiegler called in his conference the French left to 're-elaborate a political project concerning knowledge, not only to be in opposition but also to propose, elaborate, project and rebuild'.

This position has been re-iterated by the new president of the AAGT-AI, Giuseppe Longo, in a private email conversation with other members of the association. Here, he expresses the hope of inventing 'an appropriate way to interface science and socio-political action'. As he puts it: 'in these times, the effort to build scientific objectivity, for example of the ecosystem but also of the psyche, already has a political value. In fact, the critical content that is proper to scientific activity is today threatened by neo-capitalism'.[23] The new leaders of the Association thus take over from the tradition of Ars Industrialis in elaborating a new critique of science, for a science that it is not integrally submitted to the interests of capitalist development but instead succeeds in fulfilling its function to heal (*panser*) and re-vitalize societies.

The AAGT aims in this way not only to work *with* scientists but also to work *on* science, if not on its 'actors'. This goal is fundamental in the context of the Covid-19 crisis, in which some scientists contribute to discredit scientific knowledge.[24] Indeed, according to Stiegler, this crisis illuminates to the extent to which indeterminacy and uncertainty have been repressed and occluded as issues, and with them the second law of thermodynamics named entropy – a repression most acutely revealed by the astonishing absence of its mention, in French preliminary school programmes.

The school of, and for, the Thunberg generation

From the above arises the imperative to establish what Stiegler conceptualized as the 'School of [and for] the Thunberg Generation', consolidated during the first half of 2020. At first idealized as 'itinerant' between different territories, the Covid-19 outbreak compelled us to extend it to an online space: a school that is based on open-source, encrypted and non-extractive technologies and interfaces, and that consequently had to experiment alternative digital tools that are more open to online re-appropriation, more practical for the sharing of knowledge and more valorizing of deliberation, controversy and rational discussion among its members.

With climate activist and computer engineering student Basile Leretaille and designer and *chargé d'expérimentation* at the IRI, Riwad Salim, we worked on thinking and building our technological landscape, notably with our digital Mattermost forum. This forum attempts to integrate a long-going reflection on categories and meta-categories undertaken by Stiegler and the IRI, and adapt it to the specific context of the Thunberg generation's own way of sharing knowledge among each other (which Leretaille practised intimately, by being involved in the creation and organization of numerous Discord forums for the climate youth).

The structure of the AAGT-AI itself can be said to be a general experimentation of what can constitute a new way to transmit and co-build knowledge, between different generations and strata of citizens. It is a bet, made first and foremost by Stiegler, and not unlike the one made by the Internation network, that the new generation of engaged citizens can work with complicated materials, even if it requires a long time to integrate the vocabulary, generally unknown to them. Our bet is that this vocabulary and knowledge spreads in civil society and youth movements: such as the term 'entropy' that, half a year after the start of the AAGT, did already figure on demonstration banners.[25]

At last, we consider this school, before anything else, as a space in which time is invested in productive leisure and collective individuation: something which the Latin term *otium* and its Greek cousin *scholē*[26] most accurately express, both implying a certain type of collective emancipation, made possible through transindividual learning. The 'School of the Thunberg Generation' is a place where we try to re-learn how to live and contemplate; it is a space in which studious activity becomes its own end, one that frees itself from the constraints of utility.[27] In that sense, the AAGT ultimately ambitions to interrogate the role, content and methods of contemporary education in face of our critical situation, by experimenting of a school capable of transforming our 'current *state* of Anthropocene' into a *right* to the Neganthropocene.[28]

Notes

1. As Bernard Stiegler wrote it, in his suggestions for my first article on the Mediapart blog of the Association, clarifying what we mean by 'dreams', cf. Victor Chaix, 'Une plateforme de savoirs pour sortir de l'Anthropocène', *Le blog de Les amis de la génération Thunberg*, *Mediapart* (30 January 2021), available at: https://blogs.mediapart.fr/les-amis-de-la-generation-thunberg/blog/300120/une-plateforme-de-savoirs-pour-sortir-de-l-anthropocene.
2. These demands correspond, more precisely, to XR UK's demands as displayed on their website, available at: https://extinctionrebellion.uk/the-truth/demands/.
3. Cf. IPCC, *Special Report: Global Warming of 1.5°C* (2018), available at: https://www.ipcc.ch/sr15/; and IPBES, *Global Assessment Report on Biodiversity and Ecosystem Services* (2019), available at: https://ipbes.net/global-assessment.
4. See Bernard Stiegler, *Prendre Soin 1: De la jeunesse et des générations*, Paris: Flammarion, 2008.
5. More commonly referred, in the media and social demonstrations, as the 'climate generation'.
6. Often the ones that are in positions of power in society, but that are also close to leave, with retirement.
7. IRI, 'Il faut défendre Greta Thunberg', round table at *Les entretiens du nouveau monde industriel 2019. International, internation, nations, transitions: Penser les localités dans la mondialisation* (17 December 2019, Centre Pompidou, Paris), recording available at: https://enmi-conf.org/wp/enmi19/table-ronde-1/.
8. Association des Amis de la Génération Thunberg & Institut de Recherche et d'Innovation, *Rencontres préparatoires au colloque de la Sorbonne 2020* (online from 30 April 2020 to 25 June 2020), recordings available at: https://iri-ressources.org/collections/collection-50.html.
9. Victor Chaix, 'Qu'est-ce que la génération Thunberg ?', *Le blog de Les amis de la génération Thunberg*, *Mediapart* (18 June 2020), available at: https://blogs.mediapart.fr/les-amis-de-la-generation-thunberg/blog/180620/quest-ce-que-la-generation-thunberg.
10. See the heading 'Génération Covid: La jeunesse victime de la crise économique', *Le Monde* (14 June 2020), front page (paper edition).
11. *Pharmakon*-seminars, 'Exorganologie III – Remondialisation, Localités et modernité' (Paris and online, 2020), recordings available at: https://iri-ressources.org/collections/season-59.html.
12. Bernard Stiegler, *Taking Care of Youth and the Generations*, trans. Stephen Barker, Stanford: Stanford University Press, 2010.
13. See the archive of the Ars Industrialis website, available at: https://arsindustrialis.org.
14. See 'Planck's principle', in *Wikipedia*, available at: https://en.wikipedia.org/wiki/Planck%27s_principle.

15 See the YouTube channel of Ars Industrialis, available at: https://www.youtube.com/user/carolinestiegler/videos.
16 See 'Vocation de l'association des amis de la génération Thunberg', *Le blog de Les amis de la génération Thunberg, Mediapart* (19 December 2019), available at: https://blogs.mediapart.fr/les-amis-de-la-generation-thunberg/blog/040120/vocation-de-l-association-des-amis-de-la-generation-thunberg.
17 Bernard Stiegler, 'À l'école de l'*oikos*. Le nouveau devoir d'éduquer', postface in Rodrigo Arenas, Edouard Gaudo & Nathalie Laville, *Dessine-moi un avenir: Plaidoyer pour faire entrer le 21e siècle dans l'école*, Arles: Actes Sud, 2020, 107–19.
18 These 'lessons' have been centralized by Elvira Højberg, on a special and precious page of the AAGT-AI website, and are available at: https://generation-thunberg.org/pharmakon/cours/.
19 In a private email to the Internation Collective members, on Christmas, and in preparation of the press conference to be held on 10 January 2020 in Geneva to present the collective work, Stiegler ended his email with 'bonne suite des fêtes de la fin de l'ère Anthropocène': which could suggest (with some touch of humor) the historical significance, in his own eyes, of the work he directed with the Internation network. The recording of the press conference in question is available at: https://www.youtube.com/watch?v=_fmc-VPtWx8&list=-PLqiWjLu9e9QN0W2XfnBhs9KcRg_rae-bt&t=5113s.
20 See the editorial 'Le président des amis du Palais de Tokyo limogé après un appel au meurtre de Greta Thunberg', *Mediapart* (30 September 2019), available at: https://www.mediapart.fr/journal/france/300919/le-president-des-amis-du-palais-de-tokyo-limoge-apres-un-appel-au-meurtre-de-greta-thunberg?onglet=full.
21 Bernard Stiegler, 'Pourquoi et comment philosopher, aujourd'hui', *Académie d'été d'Ars Industrialis* (Épineuil-le-Fleuriel, 26 August 2011), recording available at: https://www.youtube.com/watch?v=7PC-WDHxXU0.
22 Bernard Stiegler, 'Savoirs et Politique' (Café Repaire, 9 May 2011), recording available at: https://www.youtube.com/watch?v=6oZjpUGxC3M.
23 Giuseppe Longo, in a private informal email conversation with a group of members of the AAGT-AI (3 April 21).
24 Bernard Stiegler, 'Démesure, promesses, compromis 2. Incertitude et indétermination', *Mediapart* (7 September 2020), available at: https://blogs.mediapart.fr/edition/les-invites-de-mediapart/article/070920/demesure-promesses-compromis-23-par-bernard-stiegler.
25 See the picture 'Manifestation XR sur les Champs-Élysées, 30/05/20', inserted in: Victor Chaix, 'L'entropie comme clef de lecture et de bifurcation de l'Anthropocène', *Le blog de Les amis de la génération Thunberg, Mediapart* (18 June 2020), available at: https://blogs.mediapart.fr/les-amis-de-la-generation-thunberg/blog/190420/l-entropie-comme-clef-de-lecture-et-de-bifurcation-de-l-anthropocene.
26 See the long definition of the term in the glossary of Ars Industrialis, available at: http://arsindustrialis.org/skholé.
27 See the text 'École Thunberg' on the *Generation Thunberg* website, available at: https://generation-thunberg.org/projet/ecole.
28 Many thanks to Elvira Højberg and Giuseppe Longo for the re-reading and amelioration of this article.

Chapter 16

ANOTHER SOCIAL NETWORK IS POSSIBLE!

Harry Halpin & Geert Lovink

There are always more beginnings, further back in time, parallel links, connections, tips, rumours, readings. As the eulogies for Bernard Stiegler commence, we must take care to not let his magisterial philosophical opus overshadow both the overtly technical and political aspects of his legacy; a life dedicated to the philosophy of technology that ran parallel to the invention, introduction and eventual domination of planetary life by the Internet. One connection that is all too easy to overlook is Stiegler's engagement with what is at stake in the development of digital social networks. In contrast to most other intellectuals, Stiegler engaged these networks both on the plane of philosophy and that of running code. This question of social networks has become ever more pressing as Google, Facebook, WhatsApp, Instagram, Twitter and other platforms collectively known as 'social media' rapaciously enveloped the totality of our collective existence. At the present moment, Facebook has already absorbed nearly a third of sentient life as *friends*. In 2010 when Facebook was still a new phenomenon, Stiegler presciently asked:

> But what does this term mean here, 'friends'? To what type of relation does it refer, and how do digital relational technologies implemented by social networks affect the relation known as 'friendship'?[1]

Parallel to Stiegler's work, in order to grasp what might be called the originary thinking of social networks, another origin story needs to be unconcealed: the birth of the global digital social network itself as a gesture of friendship in resistance to the shock of neoliberal capitalism. At the dawn of the Internet in 1994, from the most unexpected of locations – the mountains of the Mexican Southeast – the Ejército Zapatista de Liberación Nacional (EZLN), otherwise known as the Zapatistas, seized the city of San Cristóbal de las Casas with the slogan 'Another world is possible!' protesting against the ravages of neoliberal capitalism. For decades, the Zapatistas, usually via Subcomandante Marcos, sent out communique after communique to the world, poetically asking for solidarity. Surprisingly, the world responded, forming the alterglobalization movement that called itself a 'network of networks' – a phrase also used at the time to describe the internet.[2]

This loose network collectively decided to descend from the digital realm into the all-too-concrete streets in 1999 to organize massive protests against the World Trade Organization (WTO) in Seattle in the United States. Yet the most important moment ended up not being the press centre itself, but the creation of the Indymedia open-publishing web platform, which allowed any person to upload text and photos to the website without permission. This hybrid concept of 'tactical media' allows information to be instantly displayed to the whole world, and soon led to a network of interlinked proto-blog websites synthesized as the *indymedia. org* website. Strangely enough, the anti-capitalist Indymedia could claim to have invented the 'status update' in response to the American corporate media not covering any protests against global capitalism. There was an underappreciated feature of this original form of social media that would be key to Stiegler's analysis of new forms of social networks: the status updates were not united into a stream by virtue of their creation by the same individual, but they were rather created as a collective timeline formed for the all-too-practical shared task of supporting protests: collective techno-individuation by design. Nonetheless, despite the origins of social media as a way of bypassing censorship, the anti-authoritarian principle of open publishing was soon to be inverted, turning social media into a technology of mind control.

The profit-making potential of users forming digital social networks did not go unnoticed by Silicon Valley venture capital, and for good reason: what Pierre Musso calls *network ideology* comprises the unarticulated theoretical underpinnings of Silicon Valley. Under this ideological perspective, top-down corporate 'networks' are rolled out without a purpose except to accelerate their own growth. Musso argues that network ideology is not new, since the network concept is merely a renewal of a certain positivist philosophy of Saint-Simon from the dawn of the industrial era, when the followers of Saint-Simon imagined that a vast network would unite all of humanity, abolishing archaic national and religious boundaries via the proliferation of new industrial interconnections in the forms of canals and railways.[3] One can easily see the similarity with Facebook's mission 'to connect the world'. Two years before Indymedia began, a social networking site called SixDegrees – named after the long-standing and empirically persistent theory that each individual in the world is connected to everyone else by, at most, six degrees of separation – created a new kind of website where individuals created personal profiles and then could explicitly list their friends and send them messages. Like Indymedia, SixDegrees allowed status updates, but crucially, these status updates were bound to the individual rather than a collective group.

As the anti-globalization protest movements around Indymedia were crushed by a global wave of 'anti-terrorism' hysteria after 11 September 2001, Mark Zuckerberg launched Facebook in 2003. While SixDegrees and other simultaneous efforts like Friendster failed to achieve the exponential growth beloved by Silicon Valley venture capital, Facebook immediately began going viral. Although Zuckerberg has been thought of as a maladjusted misfit, it should not be forgotten that he was studying a dual degree in computer science and psychology. Facebook itself was based on the ultimately shrewd psychological premise that had hitherto been unexploited: everyone wants to be like – and friends with – the up-and-coming American ruling class of Harvard.

Facebook went viral first in Harvard and started spreading from one elite institution to the next. To support such hyper growth, Facebook required investment funding to scale. One investor that made millions earlier with PayPal, Peter Thiel, saw in Facebook a capitalist enterprise whose engine of representation and then imitation – *mimesis* – was theorized by his own philosophical master René Girard as the hidden truth of civilization. So it should come as no surprise that Thiel wrote the first check to Mark Zuckerberg. Thiel's reasons for doing so were atypical for investors: according to Thiel, there is an existential crisis at the heart of the West and a new techno-social *katēchon* would be needed to stave off a coming apocalypse.

In contrast to the vacuous idealism of the neo-positivists of Silicon Valley, Thiel had written at least one serious philosophical treatise, 'The Straussian Moment', in which he claimed that the liberal individualism of the *homo economicus* had reached its endpoint and would be destroyed by the more explicitly political threats from opposing civilizations, namely Islam and China.[4] Although many have claimed Thiel to be a neoreactionary, he ended his thesis on the disturbing vision that Strauss was not sufficient, citing Strauss against Strauss: 'The Straussian project sets out to preserve the *katēchon*, but instead becomes a "hastener against its will". No new Alexander is in sight to cut the Gordian knot of our age.'[5] Contra the title of his own essay, for Thiel the Straussian moment is already over. Thiel turns to Girard, as 'the new science of humanity must drive the idea of imitation, *mimesis*, much further than it has in the past', where 'it is not overly reductionist to describe human brains as gigantic imitation machines'.[6] According to Girard, there is a core problem insofar as any human 'desires being, something he himself lacks and which some other person [the "model"] seems to possess. The subject thus looks to that other person to inform him of what he should desire in order to acquire that being.'[7]

Facebook accelerates this elementary process to the speed of light, funnelling its power through a platform that Thiel calls a 'doubly mimetic' loop, where a person that broadcasts what Stiegler would call digital tertiary retentions – likes, posts, photos, video – in turn creates cascades of imitation, and so attracts ever more imitating subjects. In complexity theory this is called 'the rich get richer' without irony, and this cycle of *mimesis* unbound on a planetary scale serves to formalize the relationships of power, re-inscribing all social relationships in a template created by Silicon Valley engineers. By first seeding this global experiment in *mimesis* within the Western ruling class at Harvard and Oxford, Facebook slowly but surely absorbs every social class into its endless hall of digital mirrors, inevitably spawning an attempt at the total digitization of all human relations. With over a quarter of all sentient life being part of this empire of networks, the networked *katēchon* had arrived to reinvigorate Saint-Simon for the twenty-first century, harnessing the all-too-human desire for status.

The spread of digital social networks to countries on the periphery of the West seemed to validate simultaneously both the anti-capitalist hypothesis of Indymedia and the capitalist extremism of Thiel, as the unrestricted and uncensored spread of news and connection led to widespread leaderless insurrection in 2011 against the hopelessly corrupt pre-digital regimes of Ben Ali of Tunisia and Mubarak in Egypt during the Arab Spring by a population that seemed – at least to the West – to want to copy the forms of life of Western democracies. Facebook seemed to be the harbinger in digital form of a new networked society, a hope on the part of the newly emerging digital natives that a Western-style democracy that they desired, could be reborn in their own countries. However, things got out of hand. Western democracy did not fulfil the needs for human dignity expressed during the Arab Spring, and a wave of state and Islamic counter-revolution led these social movements to be crushed or descend into civil war. In an echo of the anarchist roots of Indymedia, these Facebook-driven protests spread even into the heart of the West with the M15-Indignados movement in Spain and Occupy Wall Street in the United States. Bernard Stiegler was concerned over the 2011 movements of Arab Spring and Occupy, not due to their lack of success, but rather their lack of theoretical development. In his Paris office, in the Institut de Recherche et d'Innovation (IRI) opposite Centre Pompidou, Stiegler asked 'What are they reading? Have they written anything?' Although Facebook seemed to fuel these revolutionary social movements, no new Marx articulated a systematic vision of a post-capitalist or radically democratic society, and even Subcomandante Marcos of the Zapatistas stuck to poetic communiques. Instead, there was only a stream of status updates, images, tweets and livestreams of endless meetings and protests. For a reason that seemed to elude the participants of these social movements, a genuine new form of society seemed stillborn.

Bernard Stiegler himself refused to use Facebook, or let anyone he organized with use Facebook due to its toxic effects. Yet contra Girard, Stiegler recognized that the young were attracted to social media not just from blind imitation, but in search for something far deeper. As put by Stiegler,

> Yes, it is the young adults who develop the social networks, and who find in these technologies a way to reconstitute what they miss so dearly: namely, a *philia*. But a young adult needs the gaze of another young adult, of a peer – and that is exactly what these networks provide.[8]

Where Thiel posits the force of blind imitation as the driver of social media addiction, Stiegler rebukes by positing the need for the fundamental feeling of love. Yet this kind of love – *philia* – goes beyond the more obvious digital reproduction of friendship (*philotēs*) and the reputation that comes from attention (*kleōs*) in terms of 'likes' on Facebook, and is rather the well-spring of solidarity between beings. As these social movements hit a dead-end in their own ability to transmute their desire for *philia* into the revolutionary thought needed for true political change by the end of 2011, Stiegler began to ponder a disturbing possibility: what if digital social networking platforms like Facebook were themselves causing this inability to think? Are live websites and apps with their never-ending updates preventing us from thinking as such? It is into this void that Bernard Stiegler voyaged in order to restore a new form of politics for the hyper-industrial era.

The core of Stiegler's philosophical approach to social networking can be found in *Technics and Time, 2: Disorientation*,[9] another one of many beginnings. While most scholars return to the first volume of *Technics and Time*, for emerging critics of social media like us, the inspiration was the second volume's chapters on 'The Genesis of Disorientation' and 'The Industrialization of Memory'. Let us recall the state of internet critique in the early 2010s, where there were surprisingly few voices coming from inside European academia, as European academia was – and to a large extent, still is – caught up in a form of 'offline romanticism' that preached rejection of technics. Instead, agenda-setting came from outside, mostly from the United States, not Europe. The main references at this time were Nicholas Carr's *What the Internet is Doing to the Brain* (2010), *#digitalvertigo* by Andrew Keen (2012), Jaron Lanier's *You are not a Gadget* (2011) and Sherry Turkle's *Alone Together* (2011). Carr shared an important reference with Stiegler, namely Maryanne Wolf's history of the reading brain, *Proust and the Squid* (2008). While these American thinkers themselves failed to generate a new philosophical and political framework, here we can see a counterforce in the making, albeit slowly. Systematizing the latent concepts in Wolf and Carr, Stiegler came to the forefront, where a special role for Derrida's pharmacology was put on the table to prevent easy and lazy solutions. The task was, and still is, to work through the digital and prevent any form of European offline romanticism that would merely preach 'digital detox' as a weekend therapy. The growing consensus in the circles around Stiegler was that a twenty-first-century philosophy of technology had to be deeply ambitious if it wanted to create new digital models for the development of genuine thought.

From early on, it was clear that a call for a collective exodus was not going to be enough; we needed to combine a radical critique of social media with the development – and roll out – of alternatives, based on entirely different values such as collaboration, community, discussion and organization instead of the collection of 'friends' and 'likes'. This alternative to Facebook would not be a platform to update deindividuated 'users' but networks for goal-driven groups and projects, hearkening back to tactical media and Indymedia. In contrast to the 'German' obsession with privacy, the idea was to develop an epistemology that would facilitate the move from 'profile' to 'project' as a core organizing principle. Preaching a healthy offline life was not enough. From early on, it was clear that we would fail to unleash critiques and alternatives unless we would get a better understanding of the underlying habits such as information addiction, short-circuited attention, distraction, which were clearly leading to a collapse of the retentions that form cultural memory.

Social networks are not just there 'to keep up with the Joneses' and lurk over the private lives of your former schoolmates and ex-partners. Socials, as they are called in some countries, still have the larger potential, promise, buried deep inside them, namely to organize social life. The way to do this was to emphasize the role of digital tools in the process of self-organization. There is more to networks than status updates, mainstream news and memes. Networks have the potential to take over crucial functions that until recently have been centralized – and controlled – by social institutions such as the village, church and party. However, the ad-driven extractive design of current social networks, shows that the latter are not interested in self-organization. The only thing these are interested in is 'engagement' (with ad-related content). They already own and administrate your network. This is where the 'organized networks' concept stepped in, networks that fight the exploitation of 'weak links' and 'likes' by 'friends' in favour of sustainable, small units, based on 'strong links' as developed by Geert Lovink and Ned Rossiter in 2005 and brought together in *Organization after Social Media*.[10] Such strong links generated not merely one-off event-based movements, but long-term commitments to networked communities.

In parallel, the Unlike Us network was initiated by our Institute of Network Cultures with Korinna Patelis. The Unlike Us network published an ambitious research programme in July 2011 with the aim to formulate a new critique of the political economy of social media, in line with Stiegler's thinking. The idea was to go beyond the 'Facebook revolution' and so critique the – at the time – hardly known data economy of advertisements, as well as to formulate alternatives to Facebook. This network of artists, designers, coders and researchers set out to deconstruct the rapid growth of the (emotional) dependency of the Silicon Valley social networks that quickly invaded the 'app space' of the newly arrived smartphones. As both the digital network and the social itself were hijacked, soon even our dearest friends had no awareness of any other way how to communicate with others except via Facebook. In response to the dangerous closing down of the networked mind and the sale of private data for political and advertising purposes, some of us signed up for the first protest-departure of 50,000 Facebook users in 2010. The premise at the time was that the hundreds of millions of Facebook users could potentially still move on to a different and less toxic platform, as flocks of users had done before, moving from homepages to blogs to Orkut and MySpace. In this way, Facebook could merely be a staging point in a longer journey towards something better … but its 'stickiness' was already being felt.

Amidst the post-financial crisis years, opposition had to be organized. At this moment, another Unlike Us conference was organized in March 2012 in Amsterdam. Everyone attuned to the looming monopoly of Facebook attended, from those who sought to nationalize Facebook like Francesca Bria to those who sought its complete destruction, like Julien Coupat. The next year, Stiegler spoke at the third Unlike Us conference in Amsterdam and outlined his philosophy of social networks as the 'new political question'.[11] Social networks are another form of grammatization where the social relationships themselves are discretized as digital representations, in the same manner as language itself discretizes phenomenological retentions into the tertiary retentions of writing 30,000 years ago. However, the transformation of social networks into digital tertiary retentions leads to disruption and control hitherto unimaginable.

Beyond Foucault's notion of biopower, Stiegler theorized that 'in the years to come, we will witness the combining of digital technologies with neuromarketing, this combination will increasingly overdetermine all other human realities, it will therefore constitute a neuropower that via digital retentional technologies will conjoin biopower and psychopower in the core of the cerebral organ itself, the brain.'[12] Capitalist firms can employ social networks to monitor behavioural expressions as 'big data'. Going beyond surveillance, to control, Stiegler later explains that

> those digital networks referred to as 'social' channel these expressions by subordinating them to mandatory protocols, to which psychic individuals bend because they are drawn to do so through what is referred to as the *network effect*, which, with the addition of social networking, becomes an automated herd effect, that is, one that is highly memetic.[13]

Self-production 'in the form of personal data' allows status updates, a form of digital tertiary retentions that 'short-circuit every process of *noetic différance*', as capitalist firms can for the first time in history '*intervene in return, and almost immediately, on psychic secondary retentions* [...] *to remotely control, one by one,* the members of a network – this is so-called "personalization"' and so 'annihilate the protentions' as to destroy the possibility of the future itself.[14]

In pharmacological fashion, where the selfsame toxin may also serve as its own cure, Stiegler posited the remarkable thesis that: 'social networks radicalise the risk of regression even further, and yet such networks also open up new possibilities, possibilities for psychical and collective individuation.'[15] Just as Stiegler's time in prison led him to reflect via note-taking on philosophy and so commence his own individuation, 'the self-profiling function could of course be an exercise in reflexivity for the person practicing it', a sort of 'auto-ethnography' or at least 'auto-sociography' that would foster individuation.[16] This insight provokes the question raised by Indymedia: what if these new kind of notes, the status update, could be re-attached to something besides individual profiles to restore the original promise of social media?

Although 'there are all kinds of socio-technological networks, and Facebook is only one instance of them', Stiegler did not believe that social networks by themselves could achieve anything much, as 'the real issue is about the arrangements of social networks with social groups (since a social network without a social group is equivalent to a mafia)'.[17] So rather than attach profiles and status updates to atomic individuals, an alternative model puts the collective group at the centre of the social network, in order to foster the transindividuation of each of its participants with each other and the group as a whole, a point brought up by younger researchers of IRI[18] in Unlike Us, and supported by Stiegler: 'As Harry Halpin and Yuk Hui have shown, this requires the implementation of a social networking technology based on the Simondonian model of individuation rather than on Moreno's sociogram.'[19]

This concept was brought down to earth through the instigation of a coding effort called simply 'Social Web', under Stiegler's direction at IRI, to build software that would serve as a support for transindividuation within a group. With the help of various interns, Harry Halpin took the lead in designing a prototype 'anti-Facebook', based on the open-source social networking codebase Crabgrass created by the anarchist collective Riseup.net,[20] one of the few remaining vestiges of the work of the founders of Indymedia. Crabgrass featured the ability to create small, invite-only groups where the entire group shared a single timeline of status updates. The group could then use a range of tools such as wikis, forums and multimedia file-sharing to accomplish their goals, whatever they may be. However, these groups were limited to those centralized on a single server. By virtue of building on the open standard XMPP,[21] a standard of the Internet Engineering Task Force (IETF)[22] for decentralized messages,[23] IRI's Social Web allowed individuals to share, discuss, edit and even index via tagging both text and multimedia across different groups spread across different servers. The eventual goal was that indexation and categorization of multimedia objects could be done collectively via the integration of tools like *Lignes de Temps*[24] for the collaborative creation and metadata-based annotation of text and video. As the hindsight given by the abrupt digitization of communication during the Covid-19 pandemic demonstrates, Stiegler had considerable foresight in envisioning a superior form of collaboration beyond the ineffective videoconferencing typified by Zoom. The goal was nothing less than the transformation of both the social web and the semantic web into what Stiegler named a *hermeneutic web* based on the archiving of digital tertiary retentions. This new hermeneutic web was envisioned as a digital agora of discussion and debate to create infinitely long protentions, dis-automating Facebook by breaking the stranglehold which the idiotic 'Like' button had on our relationships, both to media and to each other.

This 'proof-of-concept' produced considerable excitement, including Stiegler and IRI's participation in the largest study of open-source innovation in Europe and increased funding for digital social innovation.[25] Via the World Wide Web Consortium (W3C),[26] Halpin then bootstrapped a new set of protocols – the W3C Social Web standards – in order to allow diverse groups to form an open-ended 'network of networks' against the closed world of Facebook. Yet without continued financial support, Stiegler did not have the capacity to employ the talented programmers needed to mature his alternative social web against the seductively easy-to-use software of Silicon Valley. Stiegler's attempt to influence protocol design via dialogue with the inventor of the Web, Tim Berners-Lee, ground to a halt as the W3C itself became increasingly controlled by platforms like Facebook and Netflix via backing Digital Restrictions Management (DRM). This led Harry Halpin to step down from the W3C, so the W3C Social Web standards descended into bickering over technical details of the metadata formats. As the W3C was reduced to a mere mouthpiece of Silicon Valley, the W3C Web annotation standard needed to accomplish Stiegler's plan to collaboratively index content also failed to be supported by any of the major browsers. Yet this is not to say that elements of Stiegler's vision were not realized on a global scale. Parts of the W3C Social Web standards are the foundation of the interoperable Fediverse, which today forms a loose network of resistance to Facebook.[27] Lastly, the US National Security Agency (NSA) revelations caused the Social Web project at IRI to change direction dramatically. Halpin and Stiegler joined together to create a new European Commission-backed project, NEXTLEAP,[28] to create a new encrypted group-based protocol called Message Layer Security (MLS) at the IETF.[29] MLS is designed with the group as the first and foremost unit rather than the individual, and includes group-based encryption to resist surveillance. Yet Stiegler's vision of a group-based Web is perhaps even so compelling that the group-based MLS protocol is now planned for deployment not only by open-source hackers, but also Google and Facebook. However, we can hope that the possibility of transindividuation opened by these Stieglerian protocols retains its ability to subvert the hegemony of Silicon Valley's narcissistic individualism.

We will end this saga about Bernard Stiegler's foundational work on the 'social media question' with an outlook on organization. The concept of an organized network ('orgnet') is close to the way Stiegler worked with others in his various networks from Ars Industrialis to the Digital Studies Network, namely to avoid creating 'weak ties' and instead create (more or less) sustainable networks, built on 'strong ties'. This way of forming networks was originally theorized – and practised – by communities that built and maintained 'social networks' in a distributed, federated manner that followed a 'Dunbar's number', namely the controversial thesis that there are approximately one hundred and fifty productive, sustainable social connections in a community,[30] in opposition to the hyper growth model of the extractive social media platforms – which are at this point little more than advertising companies – that push individuals to try to achieve thousands, millions of 'friends' and 'followers' in order to become an 'influencer'. In opposition to the secretive economy of data mining which only results in a proletarianization of the nervous system, there remains an open conspiracy of subversive philosophical engineers that work on software alternatives and educational models that counter disorientation in order to overcome the crisis of pedagogy.

While a thorough study on Stiegler's possible – and real – contribution to organization studies ('organology') is yet outstanding, we should acknowledge his very real political activism, in this case his role in Unlike Us and his work at IRI in the period 2011–14, concerning the core concepts and concepts of future social media architectures. Stiegler called for the creation of 'politicized communities of friends in the social networks' where it should be perfectly feasible 'to go on the networks in order to counter any of these very same networks that stands in the way of their concretization as a process of psychical, technological and collective individuation' and so the circles around Stiegler were working 'to establish spaces of critique, with the aim to invent a much needed political technology'.[31] The goal was nothing less than the rebirth of autonomy as the foundation of noopolitics, a politics of knowledge capable of surpassing capitalist short-term thinking:

> The challenge of social networks is to transform the neuropower that operates on brains and on societies that have been conquered by the science-technology industry, an industry of retentions, into a noöpolitics of societies who emphasise this neuropower on themselves and by themselves, and so ensure that in the

era of digital tertiary retentions and neuroeconomics, psychic and collective individuation is politically reimagined.[32]

This noopolitics would then restore the long circuits of transindividuation through the generations that attract the young to Facebook, to make sure that alternative architectures of the social would facilitate intergenerational exchange. The challenge was – and still is – to design networks in which

> young adults are enabled to find their path toward adulthood, transforming from minors to adults in the process, a thing that has become extremely difficult in an age where adults themselves have become so dramatically infantilized.[33]

Despite its breadth of vision, Stiegler's social web has not yet managed to bring into being the digital 'republic of letters' that Stiegler hoped. Why did Stiegler's work for another social network not spread via some network effect, like Indymedia and Facebook before it? Perhaps because, as Subcomandante Marcos is claimed to have said, 'another world is possible, but only on top of the corpse of capitalism'. However, this is an all-too easy answer, as the perennial question of revolution returns: how to drive us through and beyond the mimetic social engineering embedded with Facebook, to break through the *katēchon* so that we can realize 'an unredeemed promise', namely that 'within the image of the global social network there is a picture of the possibility of a unified world', and so 'the world itself'.[34] The answer lies in the fact that Stiegler's work on the Social Web at IRI came too early, for it was only a *supplement*. The digital social network cannot serve in the place of the human social network. A technical prosthesis cannot force into being a new world, in the same way that great thinkers like Stiegler are *untimely*, coming too early for the schools of thought and social movements that only all-too-late take up the path that has been blazed for them. Being must come into its fullness in the world on its own rhythm and time, and some attempts may be like tender shoots before the last frost.

Stiegler did plant the seeds for another social network to be realized one day. An integral part of Stiegler's life, work and future legacy are the numerous groups in which he played a leading role. This spectre starts at IRI and proceeds to *Plaine Commune* and Ars Industrialis, to the *Pharmakon.fr* website and the Épineuil-le-Fleuriel summer school, and finally to the Association of the Friends of the Thunberg Generation and the Internation network. These initiatives can be seen as 'school-making' experiments, in between the non-profit model, proto-social movements, research groups and avant-garde congregations that are open to a variety of philosophic, activist and artistic misfits – as long as they were not shy to debate and think through the deep questions of this epoch without an *epokhē*, together. These clouds of activities have yet to be described in detail, and given a prominent place inside Stiegler's legacy. Here we tried to map only one episode and encourage everyone to write up similar stories and create one, two, many new organized networks.

Notes

1. Bernard Stiegler, 'Five Hundred Million Friends: The Pharmacology of Friendship', in *Umbr(a): Technology*, trans. Daniel Ross, in ed. Joel Goldbach, Buffalo: Center for the Study of Psychoanalysis and Culture, 2012, 59–75, 59.
2. Donatella Della Porta & Lorenzo Mosca, 'Global-net for global movements? A network of networks for a movement of movements', *Journal of Public Policy* (2005): 165–90.
3. Pierre Musso, 'Network Ideology: From Saint-Simonianism to the Internet', in *Pierre Musso and the Network Society*, ed. José Luís Garcia, Berlin: Springer, 2016, 19–66.
4. Peter Thiel, 'The Straussian Moment', in *Politics and the Apocalypse*, ed. Robert Hamerton-Kelly, East Lansing: Michigan University Press, 2007, 189–218.
5. Thiel, 'The Straussian Moment', 207.
6. Thiel, 'The Straussian Moment', 209.
7. René Girard, *Violence and the Sacred*, trans. Patrick Gregory, Baltimore: Johns Hopkins University Press, 1979, 146.
8. Bernard Stiegler, 'The Most Precious Good in the Era of Social Technologies', in *Unlike Us Reader: Social Media Monopolies and Their Alternatives*, trans. Patrice Riemens, in ed. Geert Lovink & Miriam Rasch, Amsterdam: Institute of Network Cultures, 2013, 16–30, 28.
9. Bernard Stiegler, *Technics and Time, 2: Disorientation*, trans. Stephen Barker, Stanford: Stanford University Press, 2009.
10. Geert Lovink & Ned Rossiter, *Organization after Social Media*, Colchester: Minor Compositions, 2018.
11. Bernard Stiegler, 'Social Networking As a Stage of Grammatization and the New Political Question' (conference presentation, dated 22 March 2013); recording available at: https://vimeo.com/63803603 (published 11 April 2013).
12. Stiegler, 'Social Networking As a Stage of Grammatization and the New Political Question'.
13. Bernard Stiegler, *Automatic Society, Volume 1: The Future of Work*, trans. Daniel Ross, Cambridge: Polity, 2016, 36.
14. Stiegler, *Automatic Society, Vol. 1*, 38.
15. Stiegler, 'Social Networking As a Stage of Grammatization and the New Political Question'.
16. Stiegler, 'The Most Precious Good in the Era of Social Technologies', 22.
17. Stiegler, 'The Most Precious Good in the Era of Social Technologies', 28.
18. Yuk Hui & Harry Halpin, 'Collective Individuation: The Future of the Social Web', in *Unlike Us Reader: Social Media Monopolies and Their Alternatives*, ed. Geert Lovink & Miriam Rasch, Amsterdam: Institute of Network Cultures, 2013, 103–16.
19. Stiegler, *Automatic Society, Vol. 1*, 142.
20. The codebase is available at: https://we.riseup.net.
21. Extensible Messaging and Presence Protocol (XMPP) is an open communication protocol for instant messaging.
22. The IETF is the leading Internet standards body. It develops open standards through open protocols.
23. Peter Saint-Andre, *IETF RFC 6120 Extensible Messaging and Presence Protocol (XMPP) Core*, published May 2011, and available at: https://datatracker.ietf.org/doc/rfc6120/.

24 The *Timelines*-tool is available at: https://www.iri.centrepompidou.fr/outils/lignes-de-temps-2/.
25 Harry Halpin & Francesca Bria, 'Crowdmapping Digital Social Innovation with Linked Data', in *The Semantic Web: Latest Advances and New Domains. Proceedings of the 12th European Semantic Web Conference*, ed. Fabien Gandon, Marta Sabou, Harald Sack, et al., Berlin: Springer, 2015, 606–20.
26 The Consortium is an international community that develops open standards to ensure the long-term growth of the Web.
27 Cf. 'Fediverse', *Wikipedia*, available at: https://en.wikipedia.org/wiki/Fediverse.
28 The project's website is located at: https://nextleap.eu.
29 The technical sheet is available at: https://datatracker.ietf.org/wg/mls/about/.
30 Patrik Lindenfors, Andreas Wartel & Johan Lind, '"Dunbar's number" deconstructed', *Biology Letters* 17/5 (5 May 2021), available at: https://royalsocietypublishing.org/doi/10.1098/rsbl.2021.0158.
31 Stiegler, 'The Most Precious Good in the Era of Social Technologies', 29.
32 Stiegler, 'Social Networking As a Stage of Grammatization and the New Political Question'.
33 Stiegler, 'The Most Precious Good in the Era of Social Technologies', 29.
34 Hui & Halpin, 'Collective Individuation: The Future of the Social Web', 116.

Part V

ECHOES: INDIVIDUATING ART

Chapter 17

MNEMOTECHNICS, ECHO AND THE DISCRETE VOICE

Mischa Twitchin

For Bernard Stiegler, '[t]hinking is a above all the history of grammatization, the history of the relations of projections and introjections between the cerebral apparatus and tertiary retentions.'[1] Exploring this history of grammatization – through the very thought of thinking, as conceived of by Bernard Stiegler – we might, then, read the three terms of my essay's title through the 'relations of projections and introjections' that make possible a culture of philosophy. This is not an abstract concern here but an attempt to reflect on these relations as they occur in an essay-film for which Stiegler generously lent his voice. This film project engages with the resonances of a phrase that itself echoes relations between the living and the dead (at least, in British cemeteries) and from which the film takes its title: 'in loving memory'.[2]

Distinct from (if not, indeed, opposed to) the Christian salvationist promise, the potential of and for 'loving memory' illuminates the threshold between religious and secular in modern funerary culture. It offers a concern with *this* world, not its transcendence in another one. The afterlife evoked is, precisely, that of memory – in an enduring split between presence and absence (haunting that between past and present), distinct from the resurrection by which this split would, supposedly, be overcome.[3] In today's social media (to which Stiegler was so critically attentive), the phantasy of undoing this split in favour of one of its terms – the present – suggests the wish-fulfilment of a virtual disembodiment. Indeed, in the technologies of audio-visual synchronization, the *pharmakon* of these media pushes memory (and its metaphors) into an oblivion beyond even that due to its commodification.[4]

Exploring the mnemotechnics (or myths) of writing, as they echo in the appeal of and to 'loving memory' (an appeal that oscillates between noun and verb), also illuminates the capacity to shift between first- and third-person narratives of the 'same' self. This capacity affirms the thought of 'thinking [as …] the history of grammatization', through the 're-actualization' of the one by the other.[5] This is manifested in the very invocation of the name 'Bernard Stiegler' here, referencing a reading of his own work made in 'loving memory' (as the subject of both the film and this essay). The capacity to shift between persons is to be understood not only descriptively, but also dynamically – whether experienced, for example, through a sense of loss or of learning, anxiety or discovery. The work of identification (not to say of idealization) with respect to others and objects – including the voice – oscillates, after all, between the divergent dynamics of both mourning and melancholia.[6]

Here the question of biography engages with the very thought of a life; not least, in and for the history of philosophy. Satirizing the conventions once supposed to distinguish between the biographies of scientists and artists – as if between the interests of the 'cognitive' and the 'moral' (themselves versions of the personal and the impersonal) – Stanislaw Lem, for example, observes (in *His Master's Voice*):

> With sufficient imagination a man could write a whole series of versions of his life; it would form a union of sets in which the facts would be the only elements in common. People, even intelligent people, who are young, and therefore inexperienced and naïve, see only cynicism in such a possibility. They are mistaken, because the problem is not moral but cognitive.[7]

With regard to philosophy, 'such a possibility' – of alternatives rather than teleologies in the thought of a life – might also be considered in the scene of Meno's aporia, where Socrates offers the myth of Persephone to evoke memory in resistance, precisely, to the lures of cynicism.[8] In Meno's example, there is a question of 'finding one's own voice' through remembering potential 'versions' of a life, as demonstrated not by addressing the politics of a slave society but in the paradoxical contingencies, singularities or 'hallucinated [...] apparitions' concerning geometrical understanding.[9] The philosophical sense of voice, however, may be understood not only reflexively as an 'internal dialogue' (a form of internal time consciousness), to which the externalization of writing attests; nor simply in the understanding of tertiary retentions made accessible through modern technologies of phonography;[10] but also in a civic, political or ethical, meaning – in the sense of 'speaking out', as Stiegler himself so clearly showed, concerning an 'ecology of spirit' (beyond simply a disciplinary conflict of the faculties).[11]

To echo the question of mnemotechnics in my title, voices are not discrete in the sense that they speak by or for themselves – such as we experience today, for instance, in the 'lifeless' sound of automated messages. Speaking, after all, entails learning to listen (just as writing requires learning to read); that is, to attend to – and to care for – the mutual implication of hearing and heard. That voices cannot be listened to without being heard is also part of 'the history of grammatization', underpinning an ethics of not wanting to speak *for* others (not least, in the maieutics of philosophy). The scenarios in which voices are heard without being listened to vary, of course, with changing technologies: from auditing a philosophy seminar (accompanied by note taking) to the practice of stenography, through to the economics of today's ubiquitous 'voice activated' data-collection devices, such as *Alexa* and *Siri*.

In the cautionary myth made famous by Ovid, Echo speaks to and of another's presence through an excluded second person address. Her tragedy suggests a thought-figure of and for an anxiety haunting the sense that one cannot speak without hearing oneself – and thus experiencing the affective disjunction (characterizing thought) between speaking and meaning. This disjunction is an instance of that 'controlled paranoia' which we call normal, when we understand it is only ourselves who hear the voices in our head. (New problems of meaning arise, of course, when we suspect that others can also hear these voices; or when we suspect that the voices come from others outside of ourselves, even when we are the only ones to hear them.) Narcissus did not care for anyone besides himself and, if he heard Echo at all, he did not listen to her. He did not hear in her voice what she expressed (distinct from what was said): that is, her speaking of love. As in the cliché of master and pupil, the possibility of 'loving' was reduced to fragments of his own speech repeated back to him – just as he saw only fleeting moments of his own image reflected from the waters in which he would drown, mistaking their surface for the depths of his own desire.

Here the curse of the gods gives mythical expression to the curse of humanity; or, at least, that of its echo chambers, which include those academic disciplines – often reconceived in today's institution-speak as sources of 'knowledge transfer' or 'knowledge exchange' – from which Stiegler wished, fundamentally, to distinguish philosophy.[12] As he writes:

> Philosophy is not just an academic discipline: it is firstly a way of living – as has always been insisted upon, and which Foucault, at the end of his life, made the motive of his research, doing so, undoubtedly, as a way of dying and of preparing to die. This is how, in particular, he interpreted *Crito* and *Phaedo* – we will come back to this question.[13]

As a practice of individuation, philosophy becomes here an *ars moriendi*; while 'coming back' to a question, articulating interpretations 'radically' (not least, when going back to Plato), would be a practice of specifying philosophy 'itself' – as a return to itself, to 'the root of questions', as Stiegler proposes.[14] For Stiegler, of course, this is pre-eminently a matter of technics, which philosophy has systematically repressed in the question of its own meaning. But it is also a question of care, of the *epimeleia heautou* in Foucault's hermeneutics of the subject; not least, when what one may 'come back to' is transformed by the possibility of phonography. The relation between care and technics (as, precisely, that of mnemo-technics) evokes the mutual implication of 'projections and introjections', expressed in the relation between *hypomnēsis* and *anamnēsis*, returning us to questions of grammatization (giving voice to relations between the living and the dead).

The transformation of thinking (of mnemotechnics) in the age of the 'discrete voice' – that is, of Echo in the age of phonography (whether analogue or digital) – is related in my film to what is manifested ('in loving memory') of care in the relation between those two fundamental aspects of the Platonic *pharmakon*: *anamnēsis* and *hypomnēsis*. As indicated by what is at the heart of both terms, the question of and for philosophy that each one bears is that of memory (*mnḗmē*), indeed, of what Stiegler calls a *noēsis* that is 'worthy of being lived in a non-inhuman way'. Concerning such *noēsis*, he writes:

> *Dasein*'s psychic retentions are made possible by tertiary retentions that are collective thanks to the very fact that they are exteriorised and spatialised. *Dasein* is thus able to share, with other psychic individuals, collective tertiary retentions that it apprehends as its *own* retentions, and which belong to *the same epoch* (and to the same 'culture') as those with whom this *Dasein shares* these retentions. From this it follows, too, that individuals of the same epoch and the same culture have, if not quite the same expectations, at least a *common horizon of the convergence of their expectations*, forming *at infinity* the common protention of a *common future* – the undetermined unity of a horizon of expectation – which is also ultimately the future of humankind, that is, of *noēsis* as worthy of being lived in a non-inhuman way.[15]

Within this horizon, philosophy – in the question of its 'non-inhuman way' – is to be remembered through a maieutics in which *anamnēsis* and *hypomnēsis* are not just disciplinary matters of philology or palaeography. They are terms to think with and not simply about, as indicating a 'belong[ing] to the same epoch' (or, indeed, the same mythology) of time and technics that is called philosophy. This 'belonging' is borne by the example of citation – the very work of relation between *anamnēsis* and *hypomnēsis* – as it appears in Stiegler's name; that is, of his own citations of Foucault and Plato (to name only these two).

Between loving wisdom (philosophy) – as with Socrates's myth of Diotima in the *Symposium* – and loving memory (*philomnesis*, perhaps), the very question of 'belong[ing] to the same epoch' (in both its literal and metaphorical audibility) resonates in the afterlife of a 'common protention of a common future'. This is encountered, for instance, in the thoughtful attention generated by a library or a cemetery (or both together, as in the case of the mausoleum of Rameses II, in which there was a library that one entered beneath the inscription, 'Healing Place of the Soul').[16] Caring for relations between living and dead memory (as, indeed, for the memory of the living and the dead) returns us to the thought of Plato in Stiegler's discussion:

> This possibility of commentary is also that of *anamnesis*: contrary to what Plato believes, *anamnesis* is far from being the contrary of *hypomnesis*; rather, the latter conditions the former. But this is an ambiguous condition, as we will see; *hypomnesis* can always prevent *anamnesis*, as Plato had feared. My dispute with Plato is not when he affirms that *hypomnesis* constitutes a danger – and I will soon show to what extent we live today in a dangerously hypomnesic era. My disagreement lies in the Platonic argument that *anamnesis* is the opposite of *hypomnesis*. I believe, contrary to Plato, that in fact *anamnesis is the good way of practising hypomnesis*. But this also means [...] that we need to overcome dialectics as such when it calls for a principle of contradiction, since hypomnesic

textuality is structurally open to a multiplicity of interpretations: such is the dia-chronicity of thought.¹⁷

In contrast, for instance, to the vaunted speed of so-called 'connections' within digital tele-communication (where connection erases attention, its value being associated with the instantaneous rather than the developmental), the memory culture that may be called 'loving' is slow enough for its medial trans-formations still to include inscription in stone (distinct simply from silicon). Stone is a medium that links us to the mnemotechnics of Palaeolithic societies, whose tele-communications with the future have endured up to the present, even if we do not now understand them. An enduring example of this 'slow' connection remains the cemetery, despite the adoption of digital memorials, about which Davide Sisto writes (addressing posthumous existence online):

> You can decide in advance to 'memorialise' your [Facebook] account following your death. Memorialising an account means that once the social network's administrators have been sent a death certificate, a copy of the obituary, or a notarised statement, the word 'Remembering' will be shown next to the person's name on their profile, followed by this phrase: 'We hope people who love (deceased user's name) will find comfort in visiting his/ her profile to remember and celebrate his/ her life.¹⁸

In the lithic-based communication of the gravestone, the transgenerational loss of memory – through its encroaching illegibility – occurs at the speed by which mould grows (as can be seen in my essay-film), rather than that by which software (and its corollary hardware) becomes obsolete. Indeed, in the latter case this potential loss, affecting the re-visiting of a virtual 'profile', now occurs at a speed that is intra-generational. By contrast to the potentially instantaneous deletion of 'Remembering' online, the growth of mould is itself a kind of organic attention, transforming this very sign of neglect with its own aesthetic qualities. Such aesthetics (i.e. in the presence of someone for whom it signifies something) changes the sense of communication (distinct from being simply entropic).

The pharmacological oscillation between legibility and illegibility, between *hypomnēsis* and *anamnēsis* (as the 'dia-chronicity of thought'), also resists an encroaching hegemony of automatization. The 'loving' promise of and for the future (as 'memory') does not consist of its predictive re-production in the present, but in the pathos of manifold differences borne by the potential of and for repetition 'itself'. In contrast to the condensations of emoji culture, with its cult of the ephemeral, 'loving memory' holds open the difference between past and future – intergenerationally – rather than collapsing them into an undifferentiated present, characteristic of the 'attention economy'.¹⁹ The monetization of attention (not least, in prevailing governmental approaches to education) – as opposed to an understanding of temporal experience being a matter of and for care – is the very definition of neoliberalism. The corrosive impact of this is especially registered in the sense of 'precarity', that punitively institutionalized corollary of 'attention deficit disorder' (where the Simondonian theme of the 'psychic and collective' is exposed to a violence masquerading as political economy).

The sense of 'loving memory', then, concerns the mutual implication of *hypomnēsis* and *anamnēsis*, in which the *pharmakon* of wanting to understand grammatization and wanting to exploit it (as a relation between the ethical and the cynical) is now a relation between reading and scanning (transforming the personal and the impersonal through so-called 'artificial intelligence'). Stiegler cites this concern in the very myth of 'human' origins (with Hesiod's legends of the brothers Prometheus and Epimetheus), evoking what he calls the 'default of origin' in anthropogenesis.[20] As Stiegler writes:

> One needs to carefully distinguish between technics as a milieu of the epiphylogenetic memory in general, and mnemotechnics. By being a technical being, the human is also a cultural being: this third technical memory that surrounds him has allowed him to accumulate an intergenerational experience that we often call 'culture'. This is why it is absurd to oppose technics and culture: technics is the condition of culture that permits its transmission. However, *our time is the era of technics, we can call it technology*: it is defined by the crisis of culture, due to its industrialisation and its submission to the imperatives of market efficiency.[21]

In the world of voice recording (Thomas Edison's waking dream), the relation between *hypomnēsis* and *anamnēsis* takes on a new implication, where the voices of the dead sound unerringly like those of the living. Hearing the dead speak in the present was the dream of Konstantin Raudive also, who anticipated that if one travelled far enough out into space – at the speed of light, beyond that of sound – one would be able to listen to those who were already long dead on earth, as if they were still present.[22] It would be as if the past returned to us from the future (which is perhaps the very promise of 'culture'). As in his experiments with what he called 'electronic voice phenomena' detected in terrestrial recordings, however, one would probably not be able to hear anything of such voices through the noise of space.

This ambiguity in distinguishing between a pattern in the reception of sound and an intention in its production – to echo again the example of Meno – is also at work in the project of a 'search for extra-terrestrial intelligence' (SETI), which has been in operation since the 1950s. Of the desire to identify an extra-terrestrial signal within cosmic noise, John Durham Peters observes that 'SETI [is] not only […] the project of understanding radio emissions from deep space but is also implicitly a sustained enquiry into our earthly dilemmas about communication.'[23] Such projected communication necessarily comes from the past and is thus 'wildly asynchronous'. Peters's comment on this is highly resonant here: 'SETI […] reveals what late nineteenth-century spiritualists knew: the unity of communication at a distance and communication with the dead.'[24]

The wished-for return *of* the dead would obviate our return *to* them, our 'coming back' to the question of their existence; that is, in the philosophical question of mnemotechnics. Travelling at a speed that would take us beyond our present finitude would remove any need for those tertiary retentions that constitute culture, whether inscribed in stone, on paper or in silicon. The sense that a recording *in* the past becomes that *of* the past would then unravel, as if it were simply in and of the present, undifferentiated from anything or anyone simultaneously – indeed, necessarily – absent. That we may be absent to ourselves as much as to others (as experienced in the relation between voice and touch, for instance) is simply ignored in the modern fantasy of the instantaneous; just as it ignores the interplay of the speed of light and that of embodied understanding, distinguishing the work of reading from the scanning of data.[25]

This interplay can be heard in the reflexive sound of a hiatus when, for example, an author in reading his own text momentarily reflects on what he has just said. In the example of 'loving memory' (as both the film and its concept), this exposes the syncopated temporality of thought caught between speaking and listening, between projection and introjection. Such an echo in the light of what is written attests, precisely, to the *anamnēsis* of *hypomnēmata*; that is, to the transformation in, and as, the experience (indeed, the life) of both thought and technics.

Concerning these ambiguities encountered between 'mnemotechnics, Echo, and the discrete voice', the Quay Brothers also offer an interesting thought experiment, featuring the wallpaper in the room where Kafka died. Recalling 'the time we went to the Kierling Sanatorium outside Vienna', they write:

> And in that very room of his death, we held up a microphone in 1990 with the sublime hope of hearing a vestige of his voice in that last conversation between Klopstock, Dora Dymant, and himself, but of course it wasn't to be and it was utterly futile. If only the tape recorder could record with a kind of carbon-dating perspective, the original wallpaper might still hold the aural DNA to the sound in the room [...] along with the functional banality of a hearse arriving the following day to take his body away.[26]

This example is, perhaps, all the more peculiar in that Kafka was virtually unable to speak in the last year of his life. He observed, for instance, in a letter to his parents, 'I am only allowed to speak in whispers, and that not too often', a situation that Kathi Diamant (who quotes this letter) describes:

> Under doctors' orders, Kafka was not supposed to speak. He communicated largely through pantomime and gestures, and by means of short, hastily written messages. About one hundred of Kafka's notes were saved and later published as *Conversation Slips*. These slips testify to Kafka's sense of humour, appreciation of nature and sensitivity towards others, which he maintained until the end, despite the agonizing pain he suffered.[27]

If 'there and then' were merely an instance of 'here and now' thought would be (perpetually) instantaneous in its expression – as if without loss, in a present that had, as it were, closed the gap with the past and caught up with the future. Without a sense of the pathos of grammatization, no one would care about the difference proposed by Plato – precisely to and for his readers – between *hypomnēsis* and *anamnēsis*, where the engagement with his thought (as Stiegler insists upon it) would become meaningless.[28] The wish to erase the difference between literally and metaphorically attending to another's 'voice' – in the relation between thought and its 'tone and prosody'[29] – is a wish to belong to another epoch than our own; one that would have no need for either philosophers or great actors. (Stiegler's own examples of cultural belonging, after all, repeatedly bring together Plato and Sarah Bernhardt!)[30]

The light needed for reading – distinct from simply scanning (or mining) data by automated processes – is also associated now with the *pharmakon* of Enlightenment; that is, with the distinction between autonomy and heteronomy in the use(s) of reason. As Stiegler proposes:

> We can no longer ignore the *irreducibly* 'pharmacological' character of writing – that is, its ambivalent character – whether writing be alphabetical or digital, etched in stone or inscribed on paper, or in silicon, or on screens of digital light. Writing, and more precisely printed writing, is the condition of 'enlightenment', and it is for this reason that Kant says that it is addressed to the 'public who reads'. But there is never light without shadow. And it is for this reason that, in 1944, Theodor Adorno and Max Horkheimer were able to perceive, in the rationalization of the world, the opposite of reason and of the *Aufklärung*.[31]

Here the question of the future concerns its own past ('in 1944', for example) – as itself a return of and to concerns with 'belonging' – not least, as a question of the 'discrete'; that is, of a grammatization that might, ostensibly, erase care from the thought of its own memory.

In the algorithmic transcription (become prescription) of thought, the maieutic process of speaking-hearing (like that of writing-reading) is supposedly reduced to a virtual abstraction of inscription – a *hypomnēsis* without *anamnēsis*. In today's automated forms of Echo, for instance, there is no pathos of the occluded second person. By contrast to an indifferent technology, discrete moments of hesitation in reading aloud, however, may still be caught in returning to a phonography that is discrete not simply in the machinic (or mathematical) sense. In the registration of a difference affecting the ostensible flow or continuity of expression, we encounter an interruption not simply of thought but as an echo of thinking. Conventionally, such hesitation is edited out of phonographic recordings, conforming to syntactical expectations of 'clarity' (or even 'transparency') in the expression of thought. Breaking the flow, as if somehow blemishing the technical image of speech in the recording, evokes the very time of the recording (as a syncopation of noun and verb). Listening becomes an echo of hearing oneself think, syncopated with the time of hearing oneself speak – or (as in 'loving memory') of hearing another's voice, syncopating the time of thought with that of hearing themselves speak.

This is what the industrialized (or, at least, the commercialized demand of 'effective communication') would erase through the synchronization of thought and expression, in the standard elisions of its grammatization.[32] In 'loving memory', however, the instance of *anamnēsis* in reading an essay again – in both French and English translation – becomes itself an occasion for thinking through the very terms of its Platonic citation. Indeed, this offers an instance of what Stiegler has to say about Plato himself:

> What allows me to dialogue with Plato today (and what made it possible for Plato to engage in a dialogue with himself, throughout the lifetime of his thought) is the written form Plato has given to his thought. And by having this written form, I can *re-actualise* this thought, or, as Husserl put it, I can *reactivate* it with a new intuition.[33]

Between the literal and the metaphorical sense of philosophical 'voice' (mnemotechnics), there appears a discrete echo of thought, oscillating between presence and absence – as also between the living and the dead, in an 'ecstasy of temporality'.[34]

In what is a potentially meaningless play on words, this syncopation in and of hearing (between saying and meaning) is what I would call an 'echophrasis' of the orthothetic. In the mnemotechnics of *anamnēsis* and *hypomnēsis*, we might say that echophrasis concerns thinking (or philosophy) 'clairaudiently' – to cite a poem by Denise Riley. Elaborating on the thought that 'death makes dead metaphor revive', Riley's poem playfully suggests that: 'Spirit as echo clowns around/ In punning repartee/ Since each word overhears itself/ Laid bare, clairaudiently.'[35] Here the three terms of my essay's title are addressed together (as at the beginning) in relation to thinking of a certain history, both in and of language – where (as Riley's poem puts it) 'each word overhears itself', in speaking of and with the dead (at least, philosophically speaking).

Notes

1   Bernard Stiegler, '*Die Aufklärung* in the Age of Philosophical Engineering', trans. Daniel Ross (2012) [13 p.], 12, available at: https://www.iri.centrepompidou.fr/wp-content/uploads/2011/02/Stiegler-The-Aufklarung-x.pdf.

2   Mischa Twitchin, *Loving Memory* (2014), available at: https://vimeo.com/105036743 (French version) and https://vimeo.com/104801436 (English version). Stiegler also generously lent his voice to another essay-film project, in six parts, called *Dahlem Dorf* (2017), looking at the closure of the Africa Galleries at the old Ethnographic Museum in Dahlem, Berlin, in preparation for their move to the new Humboldt Forum in the centre of the city, available at: https://vimeo.com/172160181 (Prologue); https://vimeo.com/179066051 (Chap. 1); https://vimeo.com/182719289 (Chap. 2); https://vimeo.com/206499116 (Chap. 3); https://vimeo.com/207908080 (Chap. 4); https://vimeo.com/177274217 (Chap. 5).

3   A different exploration of the promise to return, made between the living and the dead, is offered by Georges Saunders in his novel, *Lincoln in the Bardo* (London: Bloomsbury, 2017).
4   Bernard Stiegler, *Philosophising by Accident: Interviews with Élie During*, ed. and trans. Benoît Dillet, Edinburgh: Edinburgh University Press, 2017, 71.
5   Stiegler, *Philosophising by Accident*, 62–3.
6   As Stiegler notes, in the opening of his '*Anamnesis* and *Hypomnesis*' lecture (read by him in my *Loving Memory* film): 'We have all had the experience of misplacing a memory bearing object – a slip of paper, an annotated book, a diary, relic, or fetish, etc. We discover then that a part of ourselves (like our memory) is outside of us'. The lecture is available at: http://www.arsindustrialis.org/anamnesis-and-hypomnesis.
7   Stanislaw Lem, *His Master's Voice*, trans. Michael Kandel, Cambridge, MA: MIT Press, 2020, 6.
8   Stiegler, *Philosophising by Accident*, 41.
9   Stiegler, *Philosophising by Accident*, 63. The paradoxes of transindividuation in the 'origins of geometry' are, of course, the subject of Derrida's commentary on Husserl, the 'foundational' nature of which is precisely what is in question through its own analyses.
10 Stiegler, *Philosophising by Accident*, 69–70.
11 Stiegler, *Philosophising by Accident*, 66.
12 This concern is profoundly different, for example, from the kind of history offered recently by Stephen Gaukroger under the almost self-parodic title, 'the failures of philosophy' (Gaukroger, *The Failures of Philosophy: A Historical Essay*, Princeton: University of Princeton Press, 2020). As Stiegler observes, the power-knowledge formations of 'academic disciplines', including philosophy, 'produces a meta-noetic activity that is synchronizing and normative, defining the institutional criteria by which such activity retains its influence' (Stiegler, 'Relational Ecology and the Digital *Pharmakon*', trans. Patrick Crogan, *Culture Machine* 13 (2012): 1–19, 13).
13 Bernard Stiegler, *The Age of Disruption: Technology and Madness in Computational Capitalism*, followed by *A Conversation about Christianity with Alain Jugnon, Jean-Luc Nancy and Bernard Stiegler*, trans. Daniel Ross, Cambridge: Polity, 2019, 75.
14 Stiegler, *Philosophising by Accident*, 33.
15 Stiegler, *The Age of Disruption*, 18.
16 Diodorus, *Diodorus of Sicily*, trans. C. H. Oldfather, London: Heinemann, 1933, 173; also Mischa Twitchin, 'Loving Memory', *Performance Research* 22/1 (2017): 136–40.
17 Stiegler, *Philosophising by Accident*, 61.
18 Davide Sisto, *Online Afterlives*, Cambridge, MA: MIT, 2020, 132.
19 For a history of this phenomenon see, for example, Timothy Wu, *The Attention Merchants*, London: Atlantic Books, 2017. One might also adopt Byung-Chul Han's term 'dyschronicity' (albeit from a slightly different context) in order to describe this undifferentiated present (Han, *The Scent of Time*, trans. Daniel Steuer, Cambridge: Polity, 2017).
20 Stiegler, *Philosophising by Accident*, 48.
21 Stiegler, *Philosophising by Accident*, 58.
22 On Raudive, see https://www.subrosa.net/library/music/01-here-is-konstantin-rauvivea-part-1.mp3;jsessionid=F140B87EAF69CA784C017CA20EFDDC9F.
23 John Durham Peters, *Speaking into the Air*, Chicago: University of Chicago Press, 2000, 247.
24 Peters, *Speaking into the Air*, 248.

25 A curious meeting of these possibilities (of reading and scanning) is found in the appearance of QR codes on gravestones (Sisto, *Online Afterlives*, 166), developing the sense of the cemetery as a library in new technological ways, making a mobile palimpsest of the place of rest.
26 Quay Brothers, 'On Deciphering the Pharmacist's Prescription for Lip Reading Puppets', in *Quay Brothers: On Deciphering the Pharmacist's Prescription for Lip Reading Puppets*, ed. Ron Magliozzi, New York: MoMA, 2012, 21–7, 22.
27 Kathi Diamant, *Kafka's Last Love*, London: Vintage, 2004, 110 (and Kafka's letter quoted, 117).
28 To cite but one of innumerable instances: 'When you read the sentences from Plato's *Meno*, you do not feel that you only have an approximate image of what Plato thought: you are in an *immediate* relation to Plato's thought, and you know it intimately. You are in the very *element* of Plato's thought' (Stiegler, *Philosophising by Accident*, 60).
29 Stiegler, *Philosophising by Accident*, 61.
30 Stiegler, *Philosophising by Accident*, 64.
31 Stiegler, '*Die Aufklärung* in the Age of Philosophical Engineering', 2.
32 Stiegler distinguishes mnemotechnics from mnemotechnologies in this context (Stiegler, *Philosophising by Accident*, 62).
33 Stiegler, *Philosophising by Accident*, 61–2.
34 Stiegler, *Philosophising by Accident*, 65.
35 Denise Riley, *Selected Poems*, London: Picador, 2019, 195.

Chapter 18

BERNARD STIEGLER'S LOVE OF MUSIC

Susanna Lindberg

Bernard Stiegler was neither a musician nor a musicologist. He did not compose his philosophy in a distinctively musical manner. Yet his life twice connected with concrete musical institutions.

In his youth, he was committed to jazz. In an interview with Jean-Jacques Birgé and Jean Rochard, Stiegler tells how, prior to his prison sentence in 1978, he was a total jazz enthusiast and even had a jazz bar in Toulouse.

> 'I had created a sort of a jazz bistro in order to be able to listen to music all the time. I received musicians, I was the DJ, I thought that my job was to make others discover music – and to sell beer. I spent hours in record shops looking for good things to listen to. It was a period of my youth when I had a rigorous practice of listening. Then I stopped [...].'[1]

Later, when he had already dedicated his life to philosophy, he also collaborated with people like the philosopher-musicologist Peter Szendy, the musicologist Nicolas Donin and Laurent Bayle, who was the director of the IRCAM (Institut de Recherche et Coordination Acoustique/Musique; English: Institut for Research and Coordination in Acoustics/Music) in Paris. The IRCAM, an institution created by Pierre Boulez, is one of the world's most selective and sophisticated places for the creation of 'contemporary classical music' (this wobbly oxymoron does not really translate its French equivalent '*musique savante*', which better suits the IRCAM). In 2002, Stiegler was appointed as its director. In interviews at the time of his nomination, he recounts that he was attracted to the IRCAM because the institution was busy studying and inventing new musical technologies – not only sound syntheses but also the analysis of sound, new possibilities of digital technologies, including artificial intelligence, etc. – in a context which could foster the needs of creation rather than commerce.[2]

Are the jazz bar and the laboratory of contemporary music just biographical coincidences or symptoms of a fundamental musical experience with philosophical consequences? In reality, and although it may not be apparent at first glance, music constitutes an originary impetus for Stiegler's philosophical thinking, equal in importance to the more *visible* (precisely) cinematographic and televisual themes and – I suggest – more emancipatory than them. Music has for Stiegler a fundamental philosophical role comparable to the role of writing for Derrida and to such an extent that Stiegler even says in an interview titled 'Le circuit du désir musical' that he had wanted to study in the domain of music the question that Derrida had studied under the name of grammatology.[3]

In what follows, I will first present the 'heuristic privilege' that Stiegler accords to music on account of its 'marked instrumental nature'[4] that distinguishes it among the arts. In the domain of music, the process of instrumentalization not only began very early on but is 'originarily manifest'.[5] Music is also the best possible illustration of Stiegler's central idea of the interlacing between technics and the time of consciousness. Secondly, I will show how the question of music structures Stiegler's diagnosis of contemporary globalized society interpreted as the epoch of the industrial production of affects. Finally, I will show how music provides means of escaping the depressing standardization that is characteristic of this epoch.

Instrument and recording

The philosophical anthropology or existential analytic developed by Stiegler in *Technics and Time, 1: The Fault of Epimetheus*[6] thinks human existence (who?) in function of technical objects (what). The philosophical starting point for this work is Stiegler's objection to Heidegger's reduction of the who-what-relation to the relation between *Dasein* and its equipment (*Zeug*). One interesting counterexample to the interpretation of technical objects in terms of equipment is provided in an earlier article, 'La lutherie électronique et la main du pianiste' in which Stiegler shows that unlike equipment the musician's instrument is not a tool for … but the occasion of a specific instrumental practice in which

> the instrument is not submitted to a pre-constituted aim but on the contrary constitutes it. The aim of an instrumental practice is informed by the instrument. The pianist manifestly invents herself [*s'invente*] in her tireless practice of the instrument and this is how she invents her piano playing [*s'y invente*] endlessly.[7]

The musical instrument is not a prosthesis overcoming an organic lack either, but an instrument that is desirable because it opens a new world – although in so doing, it creates another kind of a lack, the default of instrumental skill. Its sense emerges through a 'paradox of exteriorization': the instrument seems to be an exteriorization of human interiority, but actually it only produces the interiority it presupposes. For example, the pianist expresses herself with the piano, but at the same time the piano forms the pianist's hands, memory and judgement in accordance with what the expression demands.

In the third volume of *Technics and Time*, Stiegler complements the Heideggerian starting point of the first volume by raising the Kantian and Husserlian question of the structure of consciousness. Husserl's 1905 'Lectures on the Phenomenology of the Consciousness of Internal Time'[8] is particularly important for him. In his interpretation, Husserl's example – a melody – is much more than just a possible example among others: it is the prototype of a purely temporal object that makes it possible for Husserl to articulate the structure of consciousness. This example also allows Stiegler to distance himself from Husserl by emphasizing the fundamentally technical character of all musical objects, including the melody, which turns out to be more complex than it sounds.

Stiegler returns to his reading of Husserl's 'Lectures on the Phenomenology of the Consciousness of Internal Time' in two important articles on music, 'Le circuit du désir musical' (2004) and 'Les instruments de la musique du nous' (2008).[9] In these texts he revisits his reading of Husserl's idea of consciousness as a temporal flow from the point of view of music. As consciousness is, according to Husserl, always a consciousness *of* something, the temporality of consciousness comes forth when this something is itself temporal. There is no better example of a purely temporal object than a piece of music, because it *cannot* be reduced to discursive *logos* nor to visual spatial image and yet it has a structured and finite identity. A melody is a phenomenon that only appears in its disappearance: it is pure passing that cannot be fixed without destroying its very phenomenality – and this is why its correlate, the I, also appears as a pure and incessant passing. Husserl describes this flow in terms of primary and secondary retentions, which Stiegler reinterprets and complements with his own idea of tertiary retentions.

Primary retentions are whatever consciousness retains in the flow of now, for example a resounding note. The primary retention is not a punctual impression, like an instantaneous sound frequency, but it already consists of two acts of consciousness. Firstly, in order to appear as a note, the sound must refer to other sounds in relation to which it has a musical sense (e.g. a melodic, a harmonic or a rhythmic function): an A is not really, not only, a sound of 440 Hz but a certain interval from C, D, etc. According to Husserl, the A is not A if the previous notes are not also still present in the A although they have already passed. Secondly, consciousness does not retain everything that is present in the now, but it selects meaningful elements of the now through the filter of its secondary retentions. Against Husserl, Stiegler posits that the primary retention is already selective (e.g. it retains the note but ignores the noise coming from the next room). The secondary retention is a memory of earlier primary retentions that are conserved in memory, that means not the past that is still present, but the past that is properly in the past. In the case of the melody, the secondary retentions constitute the typical features of the musical culture in which the melody is heard. For example, in Western musical culture, which is strongly marked by tonality, there is a very strong expectation that a melody in C major ends in a C-chord, instead of a chord based on a B or a D. The primary retentions filtered by secondary retentions give rise to expectations of what will follow, or protentions, like for example the expectation that a tonal melody in C major will end in C major, even though the melody has not yet come to its end. 'Music is essentially protention, that is to say, expectation.'[10] Stiegler stresses the difference between primary and secondary retentions by saying that primary retentions belong to perception while secondary retentions belong to imagination and therefore to the past proper.

Stiegler adds a new element to this Husserlian explication of the temporality of consciousness: tertiary retentions. In *Technics and Time, 3,* Stiegler shows that the intervention of the secondary retention in the act of primary retention becomes obvious only thanks to the tertiary retentions.[11] A tertiary retention is a technical object or system: a phonographic disc, a musical instrument, a score, a recording, a book, 'whatever object that serves as a support of memory that can be transmitted from one generation to another, from one individual to another, or from one culture to another'.[12] As examples of the structure that Stiegler also calls epiphylogenetic memory, these tertiary retentions 'engram' experiences on supports that have some kind of materiality (even digitalized memories are materialized). The material objectivity allows, on the one hand, for a much longer conservation of an experience than a conscious memory or even an individual experience that has fallen into oblivion. On the other hand, it allows the sharing of the experience: with an instrument, a particular *sound* can be given from a person to another; with a score, a *composition* can be shared by a whole orchestra and by many orchestras in one place and time and other places and times; with a recording, a particular *interpretation* of a given score can be shared by countless auditors. Stiegler emphasizes that our ears are not just formed by our personal listening experiences but by the entire system of tertiary retentions that make up our musical culture, for example the scores that we may or may not know (and know how to read) but that remain indispensable elements of Western musical culture.

According to Stiegler, this threefold retentional structure constitutes the musical phenomenon perceptible as a melodic/temporal flow. In the event of perceiving a melody, the temporal flow of the melody coincides with the temporal flow of consciousness. The melody is audible in function of the expectations of consciousness. For the philosopher, this reveals the structure of consciousness. But over and above this, the melody communicates its structure (as well as its unexpected features) to consciousness, for the listening consciousness adapts to the structure of the perception. In 'Les instruments de la musique du nous', Stiegler emphasizes that the adaption of consciousness to music is so total because their structure is similar.[13] This has important affective, ethical and political consequences that Stiegler has investigated in *Symbolic Misery* and in several articles.

Political music

Music's structure is not only similar to that of consciousness, it is similar to the structure of the entire soul in the Antique sense of the word. This is why music can touch, seduce, manipulate, lead and finally form the entire soul so well. This accounts for the affective force of music. Its affectivity can turn into a political force, as shown already by Plato, who imposed strict controls on music in his *Republic* in order to regulate the way in which music forms souls: like strong and courageous music forms strong and courageous souls, sly and dissolute music forms crooked citizens and should therefore be banned. Artists have always loathed the censorship that he recommends, but Stiegler follows Plato's fundamental intuition in his analysis of the political implications of the way in which music is used today in marketing.

In 'Les instruments de la musique du nous', Stiegler explores music's ethical effect and therefore its political significance with a reference to the Greek understanding of *ethos*. Continuing his analysis of Husserl but referring also to Bernard Lortat-Jacob, Stiegler stresses that in music one cannot separate the musical object from the listening subject:

> the musical object only presents itself as a subject, or more precisely as listening [*écoute*] (it is essentially listening), while listening can also be, and it very often is what gathers singular listeners into a listening 'we'. [...] I believe that this relation between the 'I' and the 'we' constitutes every musical experience: the music that resounds in my intimacy, and that can only resonate in me as my intimacy, is precisely the outside: I only hear it as something that resonates outside, as what comes from the outside to resonate in me, as what is always already outside of me and in this sense always forms the horizon of a 'we', being at the same time very intimate and radically exterior to me.
>
> I believe that this is also the structure of the ethical relation.[14]

Switching over to Gilbert Simondon's vocabulary, Stiegler continues by explaining that all psychic individuation constituting an 'I' (e.g. somebody speaks, writes, makes music) happens within a collective individuation that constitutes a 'we' (between us, we speak, write, make music in this way). Stiegler follows a motif that resurfaces regularly from Plato to Friedrich Nietzsche when he describes the individuation of a 'we' in terms of music: like music, a collective identity cannot really be explained as a discursive *logos* or as a visible image and yet it gives form to individuals. It can be a common experience, like music that attunes a public; but it can be a common experience only because – before and independently of any real experience – it contributes to the secondary and tertiary retentions that filter and orient individual experiences like a stock of pre-individual experiences. It is not by chance that the same vocabulary of apparently irrational and nevertheless effective phenomena is used to describe both music and the features and humours of collectives: affects, tonalities, harmonies and rhythms.

Ethos and music function in the same way. They constitute the most intimate intimacy of who I am, but they do this from an exteriority that is more distant than any objective, objectifiable – that is, ideal and discursive – externality, but comes from an unobjectifiable outside (*le dehors* in Maurice Blanchot's). They do not tell what to think and what to do, but they prepare the soul to think and to act or to refrain from thinking and acting. Now, if the isomorphism between music and consciousness explains the affective force of music, what constitutes the difference between ethical and totalitarian – individualizing and de-individualizing – ethico-musical experiences? In *Symbolic Misery, Volume 2: The* katastrophé *of the sensible,* Stiegler explains this difference in terms of the *lack* that translates the *'défaut qu'il faut'*. If for Plato music, together with mathematics that is consubstantial with it, is the spiritual discipline par excellence, this is because it expresses 'the transcendental affinity between consciousness and world':

> since Kant, we know that this affinity is only by default. Musical consciousness *should* be structured mathematically like the musical world, but only through a lack, and this lack is unavoidable [*c'est ce défaut-là qu'il faut*]: this lack is what makes music *ring out.* [...] The lack as that which *rings out* as *music itself*, is the power of the Siren song.[15]

To put it very bluntly, the twentieth-century totalitarian and the twenty-first-century hyper-industrial musical experiences suppress the lack by filling it overabundantly, whereas singular and therefore active musical and ethical experiences become possible when the lack is *felt* and *desire* emerges. In the modern world, the lack rises from the experience of the fissuring of the world reflected in the fissuring of consciousness that, says Stiegler, we hear in Arnold Schönberg's music.[16]

Stiegler unfolds the general structure of the musical-ethical experience by referring to its history that starts in ancient Greece, develops throughout Western history and finally becomes global. This story gets a different, brighter tonality when told as a history of different ways of 'engramming' (writing) music.

Stiegler recounts how the earliest known instruments are 45,000 years old, although our ancestors surely had musical instruments well before this.[17] After all, there do not seem to be human societies without music. The more 'primitive' the societies, the greater the function of music in the synchronization of consciousnesses appears to have been (working music, trance music, etc.; war music is also an interesting example, but Stiegler does not discuss this). Stiegler does not interpret the musical instrument as a *tool* but rather as an *'engramme'*, that is, as a 'writing' that records the same sound and makes it reproducible. It can be an instrument of collective individualization precisely because as 'writing' it makes the repetition of the same sonority possible – and each repetition is already the beginning of a rhythm. An excellent example of the capacity for collective individualization provoked by a specific instrument is the unique sound of the archaic Jewish instrument the shofar, whose uncanny capacity to evoke an entire religious culture, as described by Theodor Reik, was powerfully analysed by Philippe Lacoue-Labarthe in 'The echo of the subject'.[18]

A wholly different musical 'engram' is the musical *notation* that became the base of learned Western music since Guido D'Arezzo's invention of solmization at the turn of the first millennium[19] (although different notation technics did exist long before this). The score is a new way of producing secondary retentions: it does not conserve a sound but a *work* and it makes the distinction between composer and interpreter possible.[20]

The passage from nineteenth to twentieth century, marked by Richard Wagner, Nietzsche and Arnold Schönberg, marks a particular crisis in the European psyche in which consciousness discovers itself as incapable of harmony and unity, as reflected in the strange and dissonant music of early twentieth century[21] (and one could say that it accompanies the increasing crisis-consciousness of the nascent phenomenology). Stiegler stresses that the musical revolution that he indexes to Schönberg and Bartók (and Adorno) in particular is contemporaneous with the invention of a new technology – the invention of the *phonograph* by Thomas Edison in 1877. The phonograph, like photography, cinema, radio and television, contributes to a new synthesis of time, but Stiegler thinks music is a particularly interesting example because, unlike other industrial arts, music has a vast pre-machinic instrumental history.[22] Against this background the specificity of the new epoch shows itself distinctively. The phonograph is an unprecedented kind of a tertiary retention insofar as it, for first time, enables the absolutely identical repetition of a musical object. As Bartók already realized in 1938 and as Adorno elaborated in 1969, the mechanical reproducibility of music is a new form of musical writing that leads to a genuine revolution in our relation to music. This is why it allows the study of the changes of sensibility that accompany technological changes.[23] On the one hand, it led to a certain loss of competence because it became possible to relate to music without practising it: one could listen without eyes (to read the score) and without hands (to play it).[24] But on the other hand, the phonograph brought music to a wider public than ever before and created new practices of listening or, as Szendy puts it, 'a more thoughtful auscultation of works'.[25] Using Szendy's terms, what the phonogram makes repeatable is not only an act of hearing (*audition*) but also listening (*écoute*) – not as a simple perception but as an interpretation – that enables analytic practices of listening that can open up new interpretations, arrangements and appropriations of works.[26]

For Stiegler, the birth of jazz is an extraordinary example of what the phonograph makes possible. In 'Programmes de l'improbable, court-circuits de l'inouï', he explains how Charlie Parker really had two instruments, the saxophone *and* the phonograph with discs.[27] If the classical composer signs *compositions*, the jazz musician signs *performances*, which are 'engrammed' in discs that can then be listened again and again, and also cut in parts, recomposed, played at different volumes and speeds, etcetera: jazz music pioneers the adoption of the recording technology to constitute a new art form. Through these recordings the musician can relate to a tradition consisting of specific interpretations of a stock of standards that function as the basis for new improvisations. Such a work of art is no longer characterized by the unique aura Walter Benjamin describes in *The Work of Art in the Age of Mechanical Reproduction* nor is it a copy of a superior original. It is from the outset an interpretation, hence always already a repetition.[28] In classical music, the possibilities of conserving performances, and not just works, were discovered by Glenn Gould in particular.[29]

But on the other hand, the phonograph is also the basis for the new music industry. As Stiegler summarizes it in 'Les instruments de la musique du nous'[30], while music is always a power of synchronizing the time of consciousnesses, the new music industry extends this power even further until it becomes global:[31] music is a particularly powerful vehicle for cultural globalization because it is not limited by linguistic boundaries. The music industry also uses this power for new objectives: it standardizes behaviour in order to enhance consumption. In *Symbolic Misery, Vol. 2*, Stiegler adopts Jeremy Rifkin's description of the present hyper-industrial context:

> Jeremy Rifkin has rightly argued that we are living in the era of 'cultural capitalism'. Because it allows for the separation of producers and consumers, the machinic systemization of all forms of symbolic and sensible expressions is able to put all kinds of aesthetic spheres into the service not only of social control, but also of control societies – where it is a matter of capturing the attention of souls so as to control the behaviour of bodies, with the intention of getting them to consume goods and services.[32]

According to Stiegler, in traditional places of worship music synchronized consciousnesses in such a way that individual diachronization remained possible, but in a consumer society individual diachronization is quite simply eliminated such that the difference between 'I' and 'we' disappears and melts into the general 'they' ('*le on*', '*das Man*'). This is the present situation of aesthetic capitalism, in which all ears are submitted to formatting by industrially produced retentional dispositifs in such a way that individuals are pushed into an increasing de-individualization.[33] In the case of music, this means submitting to passively listening to (or hearing) several hours of music per day[34] without being able to select and actively produce the music one is exposed to. Even though the art world seems to provide hyper-idiomatic reactions to this, Stiegler is not sure if these constitute genuine objections to the present situation or just dialectical modulations of it.[35]

Hearing aids

According to Stiegler, in our epoch industrially produced and distributed music is one of the central modes of the aesthetic war that capitalism wages on everybody while extinguishing desire and thereby engendering a symbolic misery.[36] However, it seems to me that music is also his most developed example for new kinds of openings, because 'music is by nature addiction and desire, but positive addiction.'[37] All musical organologies are technics for synchronizing souls. Not only twentieth century totalitarianism and twenty-first-century capitalism,[38] but all musical organologies possess the capacity to turn individuals into a mindless crowd. However, all of them also open up possibilities of active listening and invention. Discovering active listening practices today does not just mean reviving amateur practices,[39] but really *thinking* through the new organological context of music.

In 'Le circuit du désir musical. L'interprète, le compositeur, l'auditeur – organes et instruments', Stiegler shows what such an active relation to music could be today. In this text, his demonstration takes the form of a deconstruction of the opposition between the composer and the interpreter in the context of Western 'learned music' (say from Monteverdi to Hector Parra and Olga Neuwirth). To start with, he emphasizes that the supposed opposition between the composer who creates and the instrumentalist who only interprets is false. Obviously, every composer works by interpreting existing secondary and tertiary retentions including the available instruments (*lutherie*) and every instrumentalist interprets the work instead of simply reproducing it. But much more essentially, both are first and foremost *musicians*, that is, persons who *hear*:

> A musician is somebody who first of all hears (*entend*), that is to say, she is primarily affected by the ear, by the ear which has also eyes and hands and a body that connects them. She cannot limit herself to calculating. She can calculate, she must calculate, but if she does this, it is only in order to give to hear what she has heard herself, and which is incalculable. This is inscribed in a circuit of desire which is constituted by a web of exclamations.[40]

What the musician hears is, of course, the stock of the pre-individual musical experiences that constitute her musical culture. But it is also all kinds of other affects that need not have anything to do with music (a person, a work of art, finally any echo of the world in which she lives). Such primary affections constitute the musician's ear. 'The composer has an auditory organ, or an auditory apparatus, which needs to translate itself into [...] the form of exclamation, that is to say, she needs to spatialize and temporalize her listening, in such a way that it becomes immediately a writing.'[41] She uses the available pre-individual stock in order to play with the auditors' horizon of expectations while trying to produce something unexpected.[42] The interpreter, for her part, does not simply *reproduce* the score but *repeats* the composer's listening, as well as her own listening that arranges and deranges the score. Without interpretation, music is only virtual, it suffers from a lack of being, it does not *exist* without this interpretative exclamation, which can also become a writing (e.g. on a disc).[43] Music is really a tissue of retentions in which the composer and the interpreter can hear a default to which both of them need to answer, but that also shows their respective defaults. Finally, the contemporary auditor, who can very well ignore how one writes and plays music, can also be an active part of the musical event. Stiegler reminds us that the auditor does not only hear a concert but also judges and interprets her act of listening and today also repeats it.[44]

Indeed, the emergence of recording technologies that first seemed to detach listeners from musical practice has led to new forms of participation in music-making and to new musical practices. Some of these are born when musical 'engrams' become means of active listening, like Miles Davis's work with discs or classical music amateur's practices in studying and annotating scores, performances and recordings. But there are also more and more musical technologies that enable an even more active relation to recordings. In 'Prologue with Chorus', Stiegler mentions musical amateurs whose practices open up new possibilities of listening (such as himself in his jazz aficionado period and today's samplers and turntablers), but also pioneers of electronic music such as Karlheinz Stockhausen and Iannis Xenakis as well as work done in avant-garde research organizations such as the IRCAM.[45]

Stiegler emphasizes that music takes place as tertiary retentions, which are not only scores, or only instruments, but also recordings, concert halls, systems of distribution and dispositifs for the formation of ears such as radios, conservatories and institutions of programming. Together they constitute the techniques that format sensibility and push towards a 'politics of the sensible life of souls and bodies'.[46] When this politics is reduced to a politics of control, it can numb sensibility. But a politics of sensibility can also become an active aesthetic judgment. One mode of such an aesthetic judgement takes form when the machinic epoch gives rise to a machinic turn in which sensibility is reconfigured, imagination is stimulated and new forms of imagination can take place. This is not only a theoretical possibility but an active practice for innumerable professional and amateur musicians who learn to play with new digital recording, sampling and distribution technologies and who even invent new digital music-making tools and instruments. They make evident a 'desire of music' that consists in a circuit of the exclamations of composers, instrumentalists, auditors and other active listeners.[47]

It seems to me that in Stiegler's thinking the desire of music that moves in this circuit has a particularly great potentiality for short-circuiting the extinction of desire: not because it should always circulate better symbols than those that push towards symbolic misery, but because whatever circulates arises from the desire of music. Stiegler says: 'Music is addiction and desire, but positive addiction [...] Real music, if you can say acting music, is a practice of putting out of control – even by the very institutions and dispositifs that make of a music a dispositif of control.'[48] It realizes a gift economy[49] that short-circuits the closed economy of capital because it constantly gives more than it has and gives what it does not even have, as Derrida said of the gift. This is why music is Stiegler's true love.

Notes

1   Stiegler's interview with Jean-Jacques Birgé & Jean Rochard (10 February 2008), 'Bernard Stiegler, La musique est la première technique du désir', *Mediapart* (7 August 2020), available at: https://blogs.mediapart.fr/jean-jacques-birge/blog/070820/bernard-stiegler-la-musique-est-la-premiere-technique-du-desir. All quotations from the original French have been translated by S.L. Stiegler describes his more distant relation to rock in a dialogue with Rodolphe Burger, 'Électricité, scène et studio', *Révolutions industrielles de la musique: Cahiers de médiologie / IRCAM* 18 (2004): 101–8.
2   The nomination is noted for example in Mark Bachaud, 'Le philosophe Bernard Stiegler est le nouveau directeur du prestigieux IRCAM. Entretien', *L'Humanité* (17 June 2002), available at: https://www.humanite.fr/node/267037, and in Bruno Serrou, 'Bernard Stiegler, nouveau directeur de l'IRCAM', *Res Musica* (4 June 2002), available at: https://www.resmusica.com/2002/06/04/festival-agora/.
3   Bernard Stiegler & Nicolas Donin, 'Le circuit du désir musical. L'interprète, le compositeur, l'auditeur – organes et instruments', *Circuit, musiques contemporaines* 15/1 (2004): 41–56, 55.

4   Bernard Stiegler, 'Prologue with Chorus. Sensibility's Machinic Turn and Music's Privilege', in Bernard Stiegler, *Symbolic Misery, Volume 2: The* katastrophé *of the sensible*, trans. Barnaby Norman, Cambridge and Malden: Polity Press, 2015, 6–20, 7. The article was first written together with Nicolas Donin, 'Le tournant machinique de la sensibilité musicale', *Révolutions industrielles de la musique: Cahiers de médiologie / IRCAM* 18 (2004): 7–17.
5   Bernard Stiegler, 'La lutherie électronique et la main du pianiste', *M/I/S (Mots / Images / Sons)* n° spécial des *Cahiers du Cirem* (1989): 229–36, 231.
6   Bernard Stiegler, *Technics and Time, 1: The Fault of Epimetheus*, trans. Richard Beardsworth & George Collins, Stanford: Stanford University Press, 1998.
7   Stiegler, 'La lutherie électronique et la main du pianiste', 229.
8   Edmund Husserl, 'Lectures on the Phenomenology of the Consciousness of Internal Time', in Edmund Husserl, *On the Phenomenology of the Consciousness of Internal Time (1893–1917)*, trans. John B. Brough, Dordrecht: Kluwer, 1990, 1–137; Bernard Stiegler, *Technics and Time, 3: Cinematic Time and the Question of Malaise*, trans. Stephen Barker, Stanford: Stanford University Press, 2011, 13–16.
9   Bernard Stiegler, 'Les instruments de la musique du nous', in Jean During, *La musique à l'esprit*, Paris: L'Harmattan, 2008, 21–34, 29–31; Stiegler, 'Le circuit du désir musical', 43–4.
10 Stiegler, 'Les instruments de la musique du nous', 31.
11 Stiegler, *Technics and Time, 3*, 18.
12 Stiegler, 'Les instruments de la musique du nous', 31.
13 Stiegler, 'Les instruments de la musique du nous', 30.
14 Stiegler, 'Les instruments de la musique du nous', 25. *Écoute* is not only hearing but listening and attentiveness (to discourses and ethical injunctions).
15 Stiegler, *Symbolic Misery, Vol. 2*, 53. Stiegler refers to Maurice Blanchot's reading of the Siren song in *Le livre à venir* (Paris: Gallimard, 1959), which I also comment on in my 'Les Filles de la Nuit', in Éric Hoppenot (ed.), *L'Œuvre du féminin dans l'écriture de Maurice Blanchot*, Paris: Éditions Complicités, 2004, 81–94.
16 Stiegler, *Symbolic Misery, Vol. 2*, 54.
17 In Birgé & Rochard, 'Bernard Stiegler, La musique est la première technique du désir', Stiegler gives the age of the oldest instruments as 40 000 years, and in 'Prologue with Chorus' as 45,000 (Stiegler, *Symbolic Misery, Vol. 2*, 8). These are the datations given by Stiegler – whether they are exact or not is immaterial. For my part, I tend to agree with Bataille: the birth of the human is the birth of art: this is another way of saying that the birth of the human is the birth of technics, as Leroi-Gourhan says. Personally, I also believe that music was there first: in humming, singing, clapping hands, stamping feet, hitting hollow trees ... organic sounds that have left no other traces than the pleasure of our body when it moulds to such rythmic and harmonic patterns.
18 Philippe Lacoue-Labarthe, 'The Echo of the Subject', in Lacoue-Labarthe, *Typography: Mimesis, Philosophy, Politics*, ed. Christopher Fynsk, with an introduction by Jacques Derrida, Cambridge and London: Harvard University Press, 1989, 139–207, 152.
19 Stiegler, 'La numérisation du son', *Communication et langages* 141 (2004): 33–41, 36. This article has also been published as 'L'armement des oreilles: devenir et avenir industriels des technologies de l'écoute', *Circuit, musiques contemporaines* 16/3 (2006): 33–42; an initial shorter version – 'La numérisation du son' – was published in *Culture et recherche* 91–92 (2002): 3–6.
20 Stiegler & Donin, 'Le circuit du désir musical', 51.

21 Stiegler, 'Les instruments de la musique du nous', 28.
22 Stiegler, *Symbolic Misery, Vol. 2*, 10.
23 Stiegler, 'La numérisation du son', 37.
24 Stiegler, 'La numérisation du son', 34.
25 Peter Szendy, 'La fabrique de l'oreille moderne', in Peter Szendy (ed.), *L'écoute*, Paris: L'Harmattan, IRCAM/Centre Pompidou, 2000, 9–49, 37. Stiegler quotes Szendy's text in 'La numérisation du son', 35.
26 Szendy, 'La fabrique de l'oreille moderne', 47, and Szendy, *Écoute. Une histoire de nos oreilles*, Paris: Minuit, 2001, 19. Like Szendy, Nicolas Donin studies the *fabrication* of the ear through different technics: the nineteenth century used scores, piano transpositions and listening guides, to which the twentieth century added all kinds of recordings, radio transmissions and now digital recordings. Both also ask how one can '*sign* one's listening', leave one's traces on a recording as on a score and use it as a source for new listenings or even new musical works. Cf. Nicolas Donin, 'Comment manipuler nos oreilles', *Révolutions industrielles de la musique: Cahiers de médiologie / IRCAM* 18 (2004): 219–28.
27 Stiegler, 'Programmes de l'improbable, court-circuits de l'inouï', *InHarmoniques* 1 (1986): 126–59.
28 Of course, the musician of the so-called oral tradition also uses different mnemotechnics (themes catalogues, standards), but the phonograph adds an unprecedented material support to these mnemotechnics. Cf. Stiegler, 'Programmes de l'improbable, court-circuits de l'inouï', 132, 143.
29 Stiegler, 'La numérisation du son', 39.
30 Stiegler, 'Les instruments de la musique du nous', 32–3.
31 See also Stiegler, 'Programmes de l'improbable, court-circuits de l'inouï', 133.
32 Stiegler, *Symbolic Misery, Vol. 2*, 12.
33 Stiegler fundamentally agrees with Adorno's interpretation of the culture industry. See in particular the introduction to Theodor Adorno, *Philosophy of New Music*, trans., ed., and with a new introduction by Robert Hullot-Kentor, Minneapolis and London: University of Minnesota Press, 2006, ix–xxx.
34 Stiegler, *Symbolic Misery, Vol. 2*, 9, 14.
35 In *Symbolic Misery, Vol. 2*, Stiegler examines Andy Warhol and Joseph Beuys in particular as artists who studied the effects of consumer society and the possibilities for new kinds of participation in contemporary consumer society. Martin Crowley claims that even Beuys's efforts have failed (Crowley, 'The Artist and the Amateur, from Misery to Invention', in Christina Howells & Gerald Moore (ed.), *Stiegler and Technics*, Edinburgh: Edinburgh University Press, 2013, 119–34, 127) while Noel Fitzpatrick claims that newer forms of participatory art might overcome Beuys' shortcomings (Fitzpatrick, 'Symbolic Misery and Aesthetics – Bernard Stiegler', *Proceedings of the European Society for Aesthetics* 6 (2014): 114–28, 126.)
36 This is the main thesis of the two volumes of *Symbolic Misery*. See also Stiegler, *Symbolic Misery, Volume 1: The Hyperindustrial Epoch*, trans. Barnaby Norman, Cambridge and Malden: Polity, 2014), vii–viii.
37 Birgé & Rochard, 'Bernard Stiegler, La musique est la première technique du désir'.
38 While Stiegler analyses the role of music in industrial capitalism, Philippe Lacoue-Labarthe analyses its role in totalitarianism and in particular Nazism in Lacoue-Labarthe, *Musica Ficta (Figures of Wagner)*, trans. Felicia McCarren, Stanford: Stanford University Press, 1994.

39 Martin Crowley stresses the role of the amateur in the taking hold of the artistic and especially musical technologies in Crowley, 'The Artist and the Amateur, from Misery to Invention', 129. Stiegler surely recognizes the importance of amateur practices, but he also emphasizes the importance of research done by artists and institutions who have more know-how as regards resisting industrial standardization.
40 Stiegler & Donin, 'Le circuit du désir musical', 42.
41 Stiegler & Donin, 'Le circuit du désir musical', 42–3.
42 Stiegler & Donin, 'Le circuit du désir musical', 48.
43 Stiegler & Donin, 'Le circuit du désir musical', 48–9.
44 Stiegler & Donin, 'Le circuit du désir musical', 52.
45 Stiegler, *Symbolic Misery, Vol. 2*, 10, 13 and 16.
46 Stiegler, *Symbolic Misery, Vol. 2*, 13.
47 Stiegler & Donin, 'Le circuit du désir musical', 46.
48 Birgé & Rochard, 'Bernard Stiegler, La musique est la première technique du désir'.
49 Stiegler & Donin, 'Le circuit du désir musical', 53; Stiegler, *Symbolic Misery, Vol. 2*, 19.

Part VI

AN UNFINISHED CONVERSATION

Chapter 19

ONTOLOGICAL DIFFERENCE, TECHNOLOGICAL *DIFFÉRANCE* AND SEMANTIC DIFFERENCE: THE PROBLEM OF A DECENTRED RECONSTRUCTION OF PHILOSOPHY AFTER 'DECONSTRUCTION'

Jean-Hugues Barthélémy

The aim of this brief paper is to clarify the nature of my critical dialogue with Bernard Stiegler during the twenty years of our friendship and intimate collaboration. This critical dialogue, arising from the encounter of two philosophical projects that were both accomplices and rivals,[1] was expressed publicly, first in my essay 'Individuation and Knowledge: The "refutation of idealism" in Simondon's Heritage in France', and then in my book *La Société de l'invention: Pour une architectonique philosophique de l'âge écologique*.[2]

I would like to refer to certain major aspects of the special relationship I have always had with him, his thought, and his work.[3] The most fundamental of these concerns the strictly architectonic question of method in philosophy, and it is this fundamental aspect that the title of my intervention emphasizes. One will see that through my critical relationship to Stiegler, it is in a sense Derrida's critical relationship to Stiegler that is expressed, albeit in a new form. I say: Derrida's critical relationship to Stiegler, and not the other way around, because my thesis is that what I have been building since 2015 under the name of 'semantic difference' is an archireflexive and post-Wittgensteinian radicalization of Derrida's questioning, to the extent that it was situated at a deeper level than Stiegler's. Geoffrey Bennington also criticizes Stiegler for not being at the level of reflexivity of Derrida's questioning. But I totally disagree with Bennington's way of talking about Stiegler: there is something in this of the contempt mere followers of one thinker can sometimes display for another thinker. Stiegler is a true thinker, and no disciple of Derrida can alter this fact. I close this parenthesis.

To outline my broad relationship to Stiegler's thought, and by way of introduction to my subject, it will be useful to go through the confrontation between my critical relationship to Simondon for thirty years and Stiegler's critical relationship to Simondon. Simondon constructed an anti-substantialist ontology of individuation which claimed to be 'first philosophy', but which at the same time explicitly claimed to be derived from schemes of physical thought.[4] This point is no longer a necessary and subtle paradox, but a real contradiction: from the moment that an alleged 'first philosophy' seeks to base itself on conceptual schemes from the sciences rather than being satisfied with a simply analogical relation, as was Kant through the idea of a 'Copernican revolution' in philosophy, it can no longer be a first philosophy, that is to say a problematics based only on itself. Now, this contradiction can be dissolved if we manage to resolve another difficulty within Simondon's discourse. Because immediately after claiming the status of first philosophy for his genetic ontology, Simondon adds: 'Unfortunately, it is impossible for the human subject to witness his own genesis, because the subject must exist in order for him to be able to think.'[5] There is indeed a new and immediate difficulty here, since the link made by Simondon between the idea of 'first philosophy' and the idea of witnessing one's own genesis means that while condemning the inclination of Husserlian phenomenologists who would bear witness to their own genesis, Simondon concedes to them that witnessing one's own genesis would be the self-knowledge in which a true first philosophy consists. Simondon also intends to propose a form of radical reflexivity, since he claims that his genetic ontology surpasses the face-to-face between the subject and the object: the object of this ontology is the process of individuation, and the knowledge of individuation is itself, he says, 'individuation of knowledge'.[6] But we can clearly see that this reflexivity does not consist in witnessing one's own genesis. Since it is 'unfortunate', as Simondon said, that the subject cannot attend its own genesis, then the knowledge of individuation that individuates itself in knowledge is a reflexivity by default.

It is by solving this difficulty that I dissolve in the same gesture the contradiction of a 'first philosophy' which would nevertheless be 'derived' from physical schemes. Indeed, in *La Societé de l'invention*, the anti-substantialist ontology of individuation is reconstructed as constituting no longer a first problematics, but rather one of the three 'second translations' of a new first problematics called 'archireflexive semantics', the two other translations being 'philosophy of economic production' and 'philosophy of axiological education'. Ontology, then, becomes the 'philosophy of ontological information', and its role is to think through the different regimes of individuation in such a way that one can ontologically account for the non-originarity of the thinking individual, just as the new first problematics had already accounted for it semantically, where the philosophizing individual had thought of himself as made by the sense(-making) that he individuates. To take account of this finitude ontologically is to leave the simple self-'knowledge' that is archireflexive semantics, with its own and radically anti-natural method, to now think of any human subject as non-originary because techno-linguistically reconstructed from the non-human animal subject, in which language and technics have not yet interpenetrated to make a cumulative history possible. This is a new, properly evolutionary, version of what Simondon called the 'transindividual', and the language/technics interface is what I call a constitutive double transcendence.

In the two volumes of his major work *Le Geste et la Parole*,[7] which appeared respectively in 1964 and 1965, and which Derrida introduced into philosophy through his remarks in *De la grammatologie*,⁸ the great French prehistorian André Leroi-Gourhan had already begun to think what he called the 'language-technics coordination' specific to *Homo*. But this coordination was thought by him as replacing, in a relatively discontinuous manner, the 'face-paw coordination' present in animals, rather than being a real interpenetration of pre-existing language and technics. Leroi-Gourhan did not analyse further the fact that human language is always, in fact, a technicized language, because it is grammaticalized, and that, conversely, human technics is always, in fact, a techno-symbolic system, because it is made of cross-references between objects. Likewise, he did not insist on animal language and animal technics. Recent neuroscientific work specifically supports this thesis of an interpenetration of pre-existing language and technics, beyond the simple 'coordination' thought by Leroi-Gourhan.⁹

Now, in the first volume of *Technics and Time*,¹⁰ where Stiegler also commented on and discussed *Le Geste et la Parole*, we already find a refoundation of Simondonian transindividual from a constitutive transcendence, since artefacts are 'crutches of the mind' or 'prostheses', in what is clearly a new understanding of this word: here the 'prosthesis' is not an artificial organ replacing a missing biological organ, but it is the external artefact in which memory is exteriorized to make reciprocally possible the construction of a true psychic interiority. Stiegler insisted here on the 'paradox', fully assumed by him, of '[having] to speak of an exteriorization without a preceding interior: the interior is *constituted* in exteriorization'.¹¹ In the first volume of *Penser l'individuation*¹² I insisted on the need to follow Stiegler rather than Simondon, insofar as in 1958, Simondon could only benefit from the first works of Leroi-Gourhan, and not yet the great work *Le Geste et la Parole*, where the human is for the first time thought of as this primate who builds his thinking interiority thanks to an exteriorization of his memory in artefacts. In his theory of the 'phases of culture', the distinctive characteristic of which was a desire to be 'genetic' without being historical, Simondon believed he could make the technics-religion pair the fundamental pair explaining the appearance of science on the one hand, and ethics on the other.¹³ With *Le Geste et la Parole* it is rather the technics-language duality that provides the basis for an otherwise properly historical process. In Simondon, language was largely absent, because it was the paradigm of his structuralist opponents, who did not contemplate a real genesis of things. It was to Stiegler's immense credit that he went further on the path opened by Leroi-Gourhan, in order to resolve certain internal tensions in Simondon's text. In *Penser l'individuation* I focused on the passages in Simondon, where the need for this artefactual refoundation of the transindividual that Stiegler had carried out made itself felt most strongly.

However, the Stieglerian refoundation of the transindividual no longer took place within a general ontogenesis which would think the physical and the living before thinking the transindividual. This was a first and really serious problem for me: Stiegler seemed to me to transform the anthropogenetic field into the first field, on the pretext that he was completely revisiting philosophical anthropology through a technogenetic questioning which, since his doctoral thesis, he saw as 'a-transcendental' rather than 'non-transcendental'.¹⁴ Moreover, the Stieglerian refoundation of the Simondonian transindividual did not consist in transforming the Leroi-Gourhanian coordination into a progressive interpenetration of language and technics. *Technē* was rather thought by Stiegler as containing within itself this artefact that would be language, and that is why *technē* was the only constitutive transcendence for the 'who' – the new name for *Dasein* as it is rethought since the first volume of *La technique et le temps*.[15] This was the second major problem in my critical relationship to what Stiegler was proposing.

These two major problems are in fact interrelated. Because the first can be divided into two sub-problems, one of which is properly architectonic, while the other concerns the question of the famous 'anthropological break' and the need to go beyond it. The second major problem, which bears on the simple or double nature of constitutive transcendence, relates precisely to this question of the anthropological break. Now, it is my new problematics of archireflexive semantics, as it is subsequently translated into a philosophy of ontological information where humans derive from animals by a progressive interpenetration of language and technics, that allows us to reconstruct philosophy after the Derridian deconstruction. And I argue that at this point Stiegler does not satisfy Derrida's radical questioning, neither about the sub-problem of philosophical architectonics, nor about the burning question of the anthropological break and its necessary supersession. To make this point, I would like to finish my brief remarks by starting again from Heideggerian ontological difference, and then recalling how my 'semantic difference' is intended to be more fundamental but also more radically anti-dogmatic, and in this way a post-Wittgensteinian heir of Derridian *différance*. I will say along the way how there is, in Stiegler, what we might call a 'technological *différance*', but whose 'a-transcendental' character does not deepen what was contained potentially in Derrida's radical questioning.

First, ontological difference, in Heidegger, is the difference between *Sein* and *Seiende*, and through it Heidegger seeks to place himself below – and beyond – all the differences that were fundamental to earlier philosophies and structured them as such. The reflexive depth of this new principial – fundamental – difference is however so difficult to explore that, in his *Nietzsche*,[16] Heidegger will criticize *Being and Time*[17] for having remained dependent of the subject-object relation by reducing the difference between *Sein* and *Seiende* to the difference between *Dasein* and other beings: in *Being and Time*, being was still thought through *Dasein*, and Heidegger affirmed, at the threshold of the work, that the ontological or 'existential' (French neologism) problem ultimately depends on the ontic or 'existentiel' (classical French linked to 'existence') level. I showed in Chapter V of *La Société de l'invention* that *Sein und Zeit*[18] suffers from another limitation, which this time will remain with Heidegger until the end. When he thinks through *Bedeutsamkeit* (*significativité*) in paragraphs 15–18 of *Sein und Zeit*, Heidegger does not exploit the archireflexive potential of the idea that appeared in §13: the idea that knowledge is only one mode within a multimodal being-in-the-world. This idea could have enabled Heidegger to question the multidimensional character of the sense(-making) of any signification – signification *as, and rather than* representation[19] – he was manipulating – or using – and which is never reducible to the sole dimension of the *ob*-ject of knowledge which would not constitute the philosophizing individual. But instead of that, instead, that is, of thinking through his own non-originarity in his relation to manipulated – or used – meanings and their multidimensional sense(-making), Heidegger, from §15, reduces the *Bedeutsamkeit* to a simple 'structure of references' between beings. The concept of *différance* in Derrida inherits this structure of references, while remaining perhaps more open to the possibility of an archireflexive radicalization as a necessary questioning on the status of the philosophizing individuals in their relation to the sense(-making) of the meanings they manipulate – or use. Here I must recall that what I refer to as significations are what have, since the *Kritik der reinen Vernunft*, been referred to as representations. The reason of this substitution is that the notion of representations does not enable us to think the multidimensionality of the sense(-making) of each object of thought, which is irreducible to the sole dimension of the *ob*-ject of knowledge.

In posing his concept of *différance*, Derrida did not simply pose the problem of writing already present in speech. It was not simply a question of going beyond linguistics by thinking of the 'system of differences' that language is as a process of self-differentiation of reference. It was also a matter of questioning thereby our relationship to the 'presence' of things, and it is at this point that my own questioning on the relativity of the simultaneity[20] between the philosophizing individuals and the sense(-making) of the meanings that they manipulate could be considered a post-Wittgensteinian radicalization of a question that is present, at least potentially, in Derrida. In any case, it is only this second aspect of his questioning that allows Derrida to escape the criticism addressed to him by Stiegler, and even to make possible a reciprocal and more fundamental criticism of Stiegler's approach. Indeed, if Derrida had been content to analyse only the writing already present in speech, then one could reduce his thought to the anthropogenetic question and criticize him for not thinking more generally of technics as constitutive of the human. With Stiegler, *différance* becomes a technological *différance*, under the names of *défaut d'origine* (originary default) and *prothéticité*. It is in fact the self-difference of the human, already thought by Sartre, who however did not understand that the being who 'has no essence' is a prosthetic being. Stiegler and Sloterdijk are the two major thinkers at the beginning of this new century who think through this prostheticity, and with Stiegler this leads to the theme of tertiary retention.

But as I said, Derrida, in posing the problem of writing, also posed the question of reference and of its own self-differentiation in the system of differences that is language. The most fundamental level of his questioning then becomes the properly post-Heideggerian problem of presence, and therefore of our relation to the sense(-making) of things that present themselves 'there in front of us'. For sense(-making) is precisely what is not there in front of us. Now, once we have understood how this fundamental problem is no longer anthropogenetic but concerns the status of the philosophizing individuals in their relation to the meanings they manipulate – or use – we can show that what is required by this fundamental level of questioning produces as consequence, at the anthropogenetic level this time, a thought of the human animal as intentionality reconstructed *via* a progressive interpenetration of technics and language. This idea of an interpenetration of technics and language is precisely what makes it possible to do away with what in Stiegler's thought looks like a residue of an anthropological break. Because in making technics that which makes the interiority of the human possible, Stiegler at the same time refused to think the non-human animal with its language and its technics, and he did not think through the process of interpenetration of language and technics through which *Homo* gradually emerged. This process, as I said above, differs from what Leroi-Gourhan had thought under the name of 'language-technics coordination', and it is also what makes it possible to find a continuity between the non-human animal and the human, as called for by Derrida when he criticized the Western philosophical tradition for not thinking the being-subject of the non-human animal.

I come, as a conclusion, to the new fundamental problematics of which all this is only a secondary ontological consequence. I said above that Heidegger, after asking the question of the 'meaning of being', forgot the question of meaning, in favour of that of being. He had certainly understood the limits of *Being and Time* and of its anchoring in *Dasein*. But he had not subjected the question of being to the archireflexive question of the meaning and the relationship of the philosophizing individuals to the sense(-making) of the meanings that they manipulate – or use. Yet *Being and Time* §13, which made knowledge a mode within multimodal being-in-the-world, opened the possibility of thinking the multidimensionality of sense(-making) of each manipulated meaning. To construct this thought is to develop a simple 'knowledge' of oneself as made by sense(-making), which cannot be reduced to the sole dimension of the *ob*-ject of knowledge. It is therefore to apply to oneself the ambition of a true thought of finitude. For it is not enough to assert that the *Dasein* thought by the philosophizing individual is finite. Neither is it enough to decentre from the point of view of ontological theses by thinking of the human from a renewed evolutionary perspective, as a techno-linguistically reconstructed primate. It is also necessary, and initially, to pose a new first problematics which is not yet ontological but semantic, and in which the philosophizing individual tries to determine the radically anti-natural method which will allow him or her to thwart the trap of human intentionality as an *ob*-jectivizing intentionality for (sense-)making. Because the philosophizing individual, by equating the manipulated meanings with their reference, reduces their sense(-making) to the sole dimension of the *ob*-ject of knowledge, whereas this sense(-making) is multidimensional, irreducible to the *ob*-ject, and it constitutes the philosophizing individual himself or herself as a non-originary subject. We therefore need to invent a methodological decentring in our relationship to the sense(-making) of the meanings that we manipulate, in order to open ourselves to a new type of principial difference: the semantic difference, which is the difference between the multidimensional sense(-making) and each of its own dimensions. Such a difference is no longer absolute but relative, and this relativity, insofar as it proceeds from an archireflexivity, must make possible what I call Philosophical Relativity.

What then is Stiegler's major contribution, since it is not a contribution concerning method in philosophy and therefore philosophical architectonics proper? As one can understand by reading *Prendre soin*,[21] Stiegler's major contribution is first the possibility of a historical psycho-sociology of the depths after the Freudian 'psychology of the depths', and that is why in chapter III of *La Société de l'invention* I used Stiegler's work, as I used that of Hartmut Rosa in chapter IV. Stiegler and Rosa are essential, beyond Freud and Marx, for thinking our time as the catastrophic result of a Western capitalist historical process. But there is another possible contribution in Stiegler's thought, and it is with this contribution that I wish to enter dialogue in my future *Critique de la raison désirante*. Because Stiegler asked himself the question of Law and of its refoundation in the ecological age, and he proposed to think of negentropy as 'value of value'. Now, even if in my thinking I separate the Law from any ethical or more generally axiological question, I nevertheless refound the Law on an economic normativity which is that of suffering needs, of which it is a question of considering the greatest possible compatibility on the scale of the biosphere, and in this the question of negentropy is not foreign to my project.

Notes

1   See my memorial on Bernard, Jean-Hugues Barthélémy, 'Au confluent du désir et de l'humain', *Un philosophe* (26 August 2020), available at: https://unphilosophe.com/2020/08/26/hommage-a-bernard-stiegler-au-confluent-du-desir-et-de-lhumain-3/.

2   Jean-Hugues Barthélémy, 'Individuation and Knowledge: The "refutation of idealism" in Simondon's Heritage in France', *SubStance* 41/3 129 (2012): 60–75; Jean-Hugues Barthélémy, *La Société de l'invention: Pour une architectonique philosophique de l'âge écologique*, Paris: Éditions Matériologiques, 2018. Of course, the book that I will devote to Stiegler's work in parallel with my future *Critique de la raison désirante* will leave much less room for critical dialogue, since it will first and foremost be a question of explaining his thought as I explained that of Gilbert Simondon, by producing an exegesis in my own manner. Here, in such a limited intervention, it did not seem to me apposite to sketch anything in terms of exegesis. For a synthesis of my exegetical work on Simondon, see Jean-Hugues Barthélémy, *Simondon*, Paris: Les Belles Lettres, 2016 (2014), and Jean-Hugues Barthélémy, *Life and Technology: An Inquiry into and beyond Simondon*, trans. Barnaby Norma, Lüneburg: Meson Press, 2015.

3   In 'Au confluent du désir et de l'humain', where I spoke of my privileged relationship with this 'big brother in philosophy' that Bernard was for me during the years 1995–2015, I recalled how our meeting in 1995 was in no sense that of a future master and a future disciple, but an encounter between a forty-three-year-old philosopher who was beginning to build his work and a young twenty-eight-year-old teacher of philosophy whose great youthful philosophical project was to be put on hold for twenty years, only to be begin in 2015, after all the prior work of controversial exegesis of Gilbert Simondon's thought had been completed. *La Société de l'invention*, published in 2018, is thus the first in a new series of works that will finally build the Philosophical Relativity which I was already planning when I met Bernard. Philosophical Relativity is the global but radically anti-dogmatic system of the individuation of sense-making,

and its name is an analogy, exactly like 'Copernican revolution' was one in Kant's *Kritik der reinen Vernunft*.

4   Gilbert Simondon, *L'individuation à la lumière des notions de forme et d'information*, Grenoble: Éditions Jérôme Millon, 2005.
5   Simondon, *L'individuation à la lumière des notions de forme et d'information*, 285.
6   Simondon, *L'individuation à la lumière des notions de forme et d'information*, 36.
7   André Leroi-Gourhan, *Le geste et la parole 1: Technique et langage*, Paris: Albin Michel, 1964; André Leroi-Gourhan, *Le geste et la parole 2: La mémoire et les rythmes*, Paris: Albin Michel, 1965.
8   Jacques Derrida, *De la grammatologie*, Paris: Les Éditions de Minuit, 1967.
9   See Barthélémy, *La Société de l'invention*, §9.
10 Bernard Stiegler, *La technique et le temps 1: La faute d'Épiméthée*, Paris: Galilée, 1994.
11 Stiegler, *La technique et le temps 1*, 152; Bernard Stiegler, *Technics and Time, 1: The fault of Epimetheus*, trans. Richard Beardsworth & George Collins, Stanford: Stanford University Press, 1998, 141 (author's emphasis).
12 Jean-Hugues Barthélémy, *Penser l'individuation: Simondon et la philosophie de la nature*, Paris: L'Harmattan, 2005.
13 Gilbert Simondon, *Du mode d'existence des objets techniques*, Paris: Aubier, 1958, 3e Partie, Chap. I.
14 Bernard Stiegler, *La faute d'Épiméthée: La technique et le temps*, Paris: École des Hautes Études en Sciences Sociales, 1993, doctoral thesis, Conclusion §3 L'a-transcendantal, 631–7. For the discussion I evoked, see Jean-Hugues Barthélémy & Vincent Bontems, 'Philosophie de la nature et artefact. La question du préindividuel', Appareil 1 (2008), available at: http://journals.openedition.org/appareil/72; DOI available at: https://doi.org/10.4000/appareil.72.
15 Bernard Stiegler, *La technique et le temps 2: La désorientation*, Paris: Galilée, 1996. For the discussion I mentioned, see Barthélémy, *La Société de l'invention*, §21, and Jean-Hugues Barthélémy, *Ego Alter: Dialogues pour l'avenir de la Terre*, Paris: Éditions Matériologiques, 2021, 137–9.
16 Martin Heidegger, *Nietzsche*, Pfullingen: Günther Neske Verlag, 1961.
17 Martin Heidegger, *Being and Time*, trans. John Macquarrie & Edward Robinson, New York: Harper and Row Publishers, 2008.
18 Martin Heidegger, *Sein und Zeit*, Fünfzehnte, an Hand der Gesamtausgabe durchgesehene Auflage mit den Randbemerkungen aus dem Handexemplar des Autors im Anhang, Tübingen: Max Niemeyer Verlag, 1979.
19 Signification is used here by way of replacement of representation, while representation coheres with the ob-jectivizing attitude that I want to overcome through its relativization.
20 This idea of the relativity of simultaneity, as it applies in philosophy to the relationship between the philosophizing individuals and the sense-making of the meanings that they manipulate, is of course that by virtue of which my primary problematics of archireflexive semantics claims the status of Philosophical Relativity or 'Einsteinian revolution' in philosophy: there is an analogy with the relativity of the simultaneity between two events of the universe in Einstein, exactly like there existed in Kant's 'Copernican revolution' an analogy between his inversion of the subject-object relation and the inversion by Copernicus of the sun-Earth relation.
21 Bernard Stiegler, *Prendre soin 1: De la jeunesse et des générations*, Paris: Flammarion, 2008.

INDEX

AAGT. *See* Association des Amis de la Génération Thunberg (AAGT)
Abgrund 29
accident/accidentality 1, 18–19, 45, 47, 53, 78–9, 93 n.89, 128, 143, 150 nn.24–5, 151 n.65, 214 nn.4–5, 214 nn.8–11, 214 n 14, 214 n.17, 214 n.20–1, 215 n.28–30, 215 n.32–4
acting out 53, 55, 60
Acting out (Stiegler) 12 n.5, 24 n.1, 25 n.11, 60 nn.4–6, 67, 74 n.31, 121 n.12, 149 nn.1–5, 149 nn.7–9, 149 nn.11–15, 149–50 nn.18–23, 150 nn.26–32, 150 n.35, 150 n.41, 150 nn.49–53, 151 n.62, 152 nn.75–7
adaptation 22, 42, 66–8, 70–1, 73 n.19, 148
addiction 19, 194–5, 224, 226
ad hominem 4
adoption 22, 43, 47, 139, 146, 169, 178, 181 n.17, 209, 223
Adorno, Theodor W. 119–20, 123 n.50, 212, 223, 228 n.33
The Age of Disruption (Stiegler) 12 n.10, 13 n.15, 13 n.29, 14 n.44, 24, 25 n.10, 25 n.12, 29, 34 nn.8–12, 34 nn.16–17, 61 n.27, 61 n.29, 62 n.35, 99, 105 n.26, 105 n.35, 113–15, 120, 121 n.19, 122 n.27, 123 n.32, 123 n.51, 133 nn.4–5, 134 n.20, 135 n.32, 135 n.38, 151 n.57, 172 n.4, 214 n.13, 214 n.15
afterlife 205, 208
agricultural revolution 5, 65–6
AI. *See* Ars Industrialis (AI)
algorithm 4–6, 54, 56, 61 n.21, 64, 101, 103, 116, 122 n.31, 162, 165 n.41, 187, 212
already-there 128, 141, 145

amateur 11, 146, 157, 160, 168, 172 n.2, 178, 186, 224–6, 228 n.35, 229 n.39
anamnēsis 8, 11, 57, 60, 140–2, 144–6, 207–13, 214 n.6
anamnesic 11, 139–41, 147, 158
condition of 140–1, 208
as frequentation 140
Anthropocene 7, 20–1, 23–4, 30, 38–9, 47, 58–9, 117, 125–6, 133 n.2, 179, 183, 186–7, 189
anthropy 39, 177
neganthropology 3, 6, 13 n.14, 28 n.28, 31, 60 n.7, 61 n.28, 61 n.31, 100–3, 108–9, 112, 117, 121 n.22, 123 n.40, 123 nn.42–3, 133 n.4
neganthropy 24, 31, 39, 67, 103–4, 145
anti-entropy 102–3, 164 n.17, 170
anxiety 44, 65, 206–7
Aristotle 5, 28–9, 34 n.5, 36, 68, 74 nn.30–1, 78–82, 85–9, 90 n.10, 90–1 nn.13–16, 91 n.22–3, 92 nn.51–54, 92 n.58, 93 n.59, 93 n.77–86, 93 n.89, 94 nn.92–3, 151 n.67
Ars Industrialis (AI) 9, 19, 22, 61 n.14, 139, 147, 167–8, 185–7, 188, 189 n.13, 189 n.15, 190 n.21, 190 n.26, 199–200
art(s) 10–11, 21, 28, 42–3, 59, 64, 139, 149 n.6, 159, 168, 177, 186, 218, 223–5, 227 n.17
artefactual benevolence 160–2
Association des Amis de la Génération Thunberg (AAGT) 183–8, 189 n.8, 190 n.16
Association of the Friends of the Thunberg Generation 9–10, 147, 183, 200
attention 1, 3–4, 49 n.33, 55–6, 58–9, 61 n.25, 82, 122 n.23, 123 n.38, 126, 139, 145, 155–6, 158, 160, 163, 170, 194–5, 208–10, 214 n.17, 224

242 Index

automatic/automating/automation/automatism 4–5, 9, 42–3, 54, 64–5, 66, 72, 116, 119, 129, 155, 158–9, 162, 168, 177, 198
Automatic Society (Stiegler) 61 n.8, 74 n.23, 75 n.57, 100, 105 nn.18–9, 105 n.22, 152 n.84, 201 nn.13–14, 201 n.19

Bateson, Gregory 160, 165 n.26, 173 n.12
Bayle, Laurent 217
becoming 3, 6, 17, 44, 46–7, 49 n.29, 54–5, 59, 68, 78, 93–4 n.89, 100–4, 108, 113, 117, 123 n.44, 128, 134 n.10, 140, 143–6, 148, 155, 161, 186
Being and Time (Heidegger) 18, 35, 77, 94 n.94, 120 n.2, 127–8, 236, 238
being/on 1, 3–5, 7, 11, 18–24, 28–30, 35–42, 47, 53–9, 63, 68, 71, 77–83, 85–9, 91 n.31, 92 n.39, 97, 99, 101, 103, 107, 110–19, 125–31, 133, 140–6, 148, 156, 160, 162, 168, 179, 185, 188, 192–4, 196, 200, 206, 208–10, 221, 224–5, 233–8
benevolent social network 161
Benjamin, Walter 223
Bennington, Geoffrey 135 n.24, 135 n.40, 233
Bifurcate: 'There is no alternative' (Stiegler and Internation Collective) 12 n.6, 152 nn.82–3, 164 n.25, 172 n.1, 180 n.1, 181 n.7, 181 nn.12–16, 181 n.23, 182 n.26
bifurcation 4, 6, 9, 24, 39, 46, 54, 60, 101, 103–4, 113, 145, 148, 159–60, 162–3, 179, 184
Bifurquer: 'Il n'y a pas d'alternative' (Stiegler and Collectif Internation) 99, 152 nn.82–3, 164 n.25, 167, 172 n.1, 180, 182 n.28
biodiversity 7, 23, 67, 69–72, 117, 178–9, 183, 189 n.3
biopower 197
The Biosphere (Vernadsky) 35–6, 48 n.1, 48 n.7
Blanchot, Maurice 6, 107–8, 118, 120 n.1, 227 n.15

blinded lucidity 119
body/bodies 11, 36, 41, 58, 64, 88–90, 92 n.53, 93 n.89, 101, 104, 110, 181 n.4, 201 n.22, 211, 224–5, 226, 227 n.17
Boltzmann, Ludwig 3, 36, 40–7, 48 n.18, 49 n.28, 49 n.30, 102
boomers 184–5
Boulez, Pierre 217

calculation 35, 63–5, 68–72, 99, 102, 126, 128, 132, 147–8, 161–2
Canguilhem, Georges 40, 48 n.16, 66–7, 73 n.20, 80, 93 n.89, 162, 163 n.5, 165 n.38
capitalism 1–3, 5, 13 n.15, 19, 22–3, 25 n.10, 34 n.8, 38, 54, 56, 61 n.26, 64, 69, 73 n.2, 99, 101, 105 n.26, 106 n.40, 121 n.19, 128 n.38, 130, 133 n.4, 151 n.57, 168, 172 n.4, 187, 191–2, 200, 214 n.13, 224
care 3–4, 11, 23, 27, 36, 38, 53–60, 100, 104, 114–20, 121 n.11, 121 n.18, 122 n.23, 123 nn.37–8, 133 n.1, 139–40, 145, 155, 159–63, 171–2, 177, 206–7, 209
carelessness 29, 38, 117, 120
Carr, Nicholas 195
category/categories 54–5, 57–9, 62 n.37, 157–61, 184, 188, 198
change 1, 3, 6, 19, 38, 56, 65, 67, 72, 100, 107–10, 112, 114, 118, 143, 147, 171, 194, 199
commerce 176, 217
community 11, 17, 79, 81–2, 135 n.43, 151 n.59, 151 n.67, 178, 195, 199, 202 n.25
computational capitalism 5, 13 n.15, 25 n.10, 34 n.8, 54, 56, 61 n.27, 64, 105 n.26, 121 n.19, 133 n.4, 151 n.57, 172 n.4, 214 n.13
The Concept of Time (Heidegger) 127–8, 134 n.13
consistence 92 n.39, 149 n.6
consumerism 19, 168, 177, 185, 224, 228 n.35

Index

contribution 1, 11–12, 19, 77, 93 n.89, 159, 161–3, 164 n.25, 165 n.39, 167, 184, 199, 238–9
    contributory 9, 22, 139, 146–8, 155, 158, 160, 162–3, 164 n.25, 167, 172, 177–80, 181 n.12, 181 nn.15–16, 181 n.23
contributive learning programme/ contributive learning territory project 9, 164, 168
Copernican revolution 234, 240 n.3, 240 n.20
Covid-19 5, 17, 19, 38–9, 68–71, 74 n.47, 74–5 n.49, 75 n.52, 146, 162, 172, 184, 188, 198
crisis 2, 21, 23, 71, 112–13, 126, 143, 147–8, 162, 188, 193, 196, 199, 210, 222–3
criterion/criteria 23, 41–3, 82–3, 85, 158, 168, 214 n.12
Crowley, Martin 75 n.56, 228 n.35, 229 n.39
curtailment 35–7

Dasein 18, 21–2, 35, 91 n.31, 104, 115, 128, 208, 218, 236, 238
data 64, 139, 147–8, 156–7, 160, 162, 196–7, 199, 202 n.25, 206, 211–12
datability 127, 128–9
De Anima (Aristotle) 74 nn.30–1, 79–81, 86, 88, 90 n.10, 91 n.23, 92 nn.51–4, 93 n.59, 93 n.77–86, 93 n.89, 94 n.93
death 4, 6, 10, 17, 27–8, 30, 31–3, 45–6, 60, 63, 65, 77–8, 84, 86–7, 89, 91 n.31, 98, 103–4, 127, 130, 144–5, 149 n.6, 151 n.62, 156, 184, 209, 211, 213
    death of God 24, 32
Decadence of Industrial Democraties, The (Stiegler) 13 nn.16–17, 13 n.17, 14 n.39, 67, 74 nn.30–1, 74 n.31, 92 n.47, 92 nn.49–50, 92 nn.55–7, 93 nn.61–3, 93 nn.65–7, 93 nn.69–70, 93 nn.74–5, 93 nn.87–9, 94 n.90, 151 n.67, 153 n.98
deconstruction 79, 81, 83–4, 87, 94 n.93, 112–13, 118, 121 n.17, 123 n.38, 129, 224, 236

default 1, 3–6, 10–11, 21, 27, 29–30, 32–3, 53, 77–88, 131, 141, 143–5, 158, 210, 218, 222, 225, 234, 237
    by default 4, 32–3, 140, 144, 222, 234
    default of origin 27, 29–30, 33, 143, 210
    necessary default 4, 29–30, 53, 77–9
défaut 31, 54
    défaut qu'il faut 28–9, 77, 158, 222
deferral 6, 65, 67, 101–2, 105 n.29, 126
delay 2, 36, 56, 102, 109, 125, 160–1, 168
Deleuze, Gilles 63–4, 73 n.1, 87–8, 165 n.35
déphasage 54
Derrida, Jacques 2, 21, 32, 49 n.33, 53, 77–80, 83, 98, 109, 115, 118, 119, 121 n.23, 123 n.38, 129–30, 132–3, 133 n.3, 176, 195, 218, 226, 233–8
desire 3, 5, 8, 10, 23, 29, 53, 55–8, 63, 67, 73 n. 2, 80, 82–3, 85, 87–8, 128, 130, 135 n.43, 146, 155, 158, 160, 162, 187, 193–4, 207, 210, 222, 224–6, 235
différance 6–7, 39, 53, 59, 101–4, 105 nn.30–1, 115–18, 126–32, 135 n.28, 135 n.33, 136 n.46, 145, 158, 197, 236–7
Digital Restrictions Management (DRM) 198
digital studies program 161
disaster 30, 36, 70, 72, 168
    disastrous 28–9, 31, 69, 115, 119
Disorientation (Stiegler) 13 n.22, 22, 90 n.12, 105 n.10, 110, 112, 121 n.7, 130, 135 n.33, 149 n.17, 195, 199, 201 n.9
dispositif/dispositive 165 n.30, 224, 226
disruption 7, 9, 22–4, 32, 59, 109–10, 114–20, 169–71, 179, 196
Donin, Nicolas 157, 217, 228 n.26
Duino Elegies (Rilke) 45, 157, 217, 226 n.3, 226 n.4, 228 n.26
Dunbar's number 199, 202 n.30

echoes (echo) 10–11, 32, 81, 85, 194, 205–7, 210–13, 222, 225
economy 9, 19, 22–3, 57–8, 99, 147–9, 155, 158, 162, 169–70, 172, 179–80, 181 n.8, 185, 196, 199, 209–10, 226
    attention 209

contributive 9, 162, 169–70
gift 226
libidinal 22–3, 40, 63, 158
political 22, 57, 99, 149, 196, 210
spiritual 57–8
ecosystem 9, 35, 40, 46, 67, 171, 183, 187, 189
Edison, Thomas 210, 223
eidos 80, 83–4
Einstein, Albert 175, 180 n.3, 240 n.20
Ejército Zapatista de Liberación Nacional (EZLN) 191
endosomatic 36, 39–42, 93–4 n.89
end(s) 9, 11, 20, 28, 31–2, 35, 37, 56–7, 59, 70, 88, 90, 108–9, 118, 120, 121 n.18, 123 n.51, 126, 129, 132, 143, 145–6, 169–70, 172, 175, 188, 194, 199, 207, 211, 219, 236
Entropocene 7, 23, 106 n.40, 109, 112, 114–19, 173 n.5, 179–80, 182 n.26
entropy 3, 6, 22, 24, 32, 36, 39–41, 43, 47–8, 67, 89–90, 97–104, 117, 125, 156, 158, 182 n.26, 188
  entropology 6, 100
  negative 36, 41, 45, 97, 102, 105 n.29, 119–20, 123 n.50
  negentropy 4–7, 22–4, 31, 45, 57, 71, 97–103, 117, 145, 156, 158–9, 163, 187, 239
Epimetheus 78, 111, 136 n.46, 210
epiphylogenesis 169–70, 210, 220
epistēmē 3, 62 n.34, 80–1, 92 n.39, 106 n.40
epochal 6–7, 53, 57–8, 60, 107–18, 121 n.11, 126, 129–30, 132
  doubling 45, 111, 113
  epochality 6–7, 108–9, 111–14, 123 n.44, 126–7, 129–31, 135 n.41
  redoubling 7, 110–17, 121 n.15, 130
epochē 7, 53, 85, 87–8, 109–12, 114, 117
epokhal redoubling 112–15, 117, 121 n.15
epokhē 109–13, 120, 121 n.11, 121 n.14, 200
essence 3, 65, 83–4, 94 n.93, 127, 128, 129, 237
existence 1, 4, 10, 21, 29–30, 36, 40–1, 46, 54–5, 58–9, 64, 66–7, 82–4, 86–7, 89, 114, 119, 148, 191, 209–10, 218, 236

exorganisms 39–40, 104, 177
exosomatization 7, 36, 40, 103, 109, 113–14
  exosomatic 36–42, 45, 47, 63, 93–4 n.89, 115–16, 121 n.16, 156, 177
experience 3–4, 10, 27–8, 32–3, 63, 77, 79–83, 98–9, 111, 115, 120, 122 n.29, 122 n.31, 130, 139, 144, 155, 160, 169, 171, 206–11, 220–2
Extensible Messaging and Presence Protocol (XMPP) 198, 201 n.21, 201 n.23
exteriorization 21, 57, 109, 141, 143, 218, 235
Extinction Rebellion 20, 183

Facebook 10, 17, 161, 186, 191–200, 209
fact 10, 32, 36–8, 40, 42–3, 45, 47, 78, 80–1, 83, 86, 91 nn.31–2, 93–4 n.89, 110, 111, 122 n.31, 128, 132, 145, 187, 151 n.67, 208, 233, 235–7
fall 40, 43, 45, 56, 168
fault 27, 31, 78, 111
Floyd generation 185
flux 83, 92 n.39
Fridays For Future 183
funeste 28–30
future, the/futurity 4–5, 9, 21–2, 30, 53, 56, 60, 63–4, 72, 97–9, 101–2, 104, 128, 148, 197, 208–10, 212

Generation Z 185
genius 18, 28–30, 46, 144
Geschlecht 7, 109, 129–32, 134–5 n.24, 135 n.24, 135 n.43
Gestell 7, 112, 115–16, 118, 122 n.29, 122 n.31, 127, 129, 131, 134 n.21
Girard, René 10, 193–4, 201
globalization 176–7, 180, 223
grammatization 90, 108–9, 122 n.31, 156–8, 196, 201 n.11, 201 n.14, 202 n.31, 205–7, 210, 212–13, 235
grammatology 118, 123 n.47, 218

Hegel, Georg W. F. 4, 21, 31–2, 34 nn.13–15, 78, 80, 85
Heidegger, Martin 1, 3, 7, 18–19, 21–3, 29, 35, 37, 41, 77, 94 n.94, 98, 102, 104, 110, 113, 115, 118–19, 122 n.29, 123 nn.37–8, 125–32, 134 n.21, 218, 236–8

hermeneutic web 198
heterochronicity 162
His Master's Voice (Lem) 206, 214 n.7
hominization 4, 36–7
Homo sapiens 24, 37
Husserl, Edmund 81, 109–11, 218–21, 227 n.6, 227 n.8, 234
hypomnēsis 8, 11, 140–2, 145–6, 207–10, 212–13, 214 n.5
   hypomnesic 140, 142–3, 147, 149 n.10, 156, 158, 180, 208

idea 68, 72, 78, 83–6, 89, 91 n.37, 112, 161, 183, 186, 193, 196, 218–19, 234, 236, 238
identity 80, 86, 89, 112, 176, 219, 221
idiocy 2, 29, 78, 81–2, 90 n.6
idios 5, 14 n.37, 79, 81, 87, 93 n.89, 151 n.67, 163 n.1
idiot 28–30, 79
idiotext 5–6, 14 n.37, 78–80, 83–5, 158
IETF. *See* Internet Engineering Task Force (IETF)
incalculable 7, 9, 24, 63, 91 n.31, 103, 126–9, 131, 145, 155, 158, 161–2, 170, 225
Incipit (Blanchot) 107
individuation 1, 8, 10, 14 n.37, 41, 44, 46–7, 54–5, 57–8, 72, 80, 99, 110, 115, 117, 130, 139–40, 142–8, 150 n.36, 157–8, 165 n.37, 178, 181 n.17, 188, 192, 197, 199–200, 207, 221, 233–4, 239 n.3
industry 11, 114, 184, 186, 199, 223–4
Institute for Research and Innovation (IRI) 8, 10, 18, 147, 156–7, 161–2, 164 n.18, 165 n.36, 170–1, 184, 188, 189 n.7, 194, 197–200
Institute for Science in Society (ISiS) 2
Institut for Research and Coordination in Acoustics/Music (IRCAM) 157, 217, 226, 226 nn.1–2, 227 n.4, 228 n.25, 26 n.26
instrument(s) 8, 35, 37, 43, 86–7, 89, 102, 111, 143, 218–20, 222–6, 227 n.17
intelligence 23–4, 142–7, 181 n.12, 210, 217
intergenerational transmission 142, 185–6
Intergovernmental Panel on Climate Change (IPCC) 183, 189 n.3

Intergovernmental Science-Policy Platform on Biodiversity and Ecosystem Services (IPBES) 183, 189 n.3
intermittence 4–6, 8–9, 54, 57, 60, 64, 66–72, 119, 155–60, 162
intermittent society 65–8
internation 9, 20, 22, 175–80, 180 n.1, 181 n.4, 183, 188
   academic internation 178
   internation collective 9, 147, 179–80
   internation network 188, 190, 200
International Commission of Intellectual Cooperation 175
Internet Engineering Task Force (IETF) 198–9, 201 n.22
interscience 9, 175, 177–80, 181 n.6, 181 n.18
invention 8–10, 27–9, 54–5, 57–9, 67, 100, 141, 143, 145–6, 148, 169, 191, 222–4, 228 n.35, 229 n.39
investment 58–9, 98, 165 n.29, 170, 193
IRCAM. *See* Institut for Research and Coordination in Acoustics/Music (IRCAM)
IRI. *See* Institute for Research and Innovation (IRI)

Kant, Immanuel 10, 56–7, 82–5, 91 n.22, 91 n.29, 91 nn.35–8, 92 nn.41–5, 159, 222, 234
katēchon 100–2, 104, 193, 200

La Société de l'invention (Barthélémy) 233–4, 236, 239, 239 n.3
Leiden University Centre for Continental Philosophy (LCCP) 2
Lem, Stanislaw 206, 214 n.7
Leroi-Gourhan, André 3, 12, 21, 81, 111, 134 n.10, 163 n.5, 227 n.17, 235–6, 238, 240 n.7
Le Geste et la Parole 234–5, 240 n.7
life 1–7, 23–4, 27, 35–7, 39–47, 53–9, 63–72, 78–9, 84–7, 89, 95, 97–9, 101–4, 108, 113, 118, 120, 126–8, 139–46, 148, 159, 168, 171, 177, 191, 193–200, 206, 211–13, 217, 226
life of the mind 23, 64, 139, 142

246 Index

local 9, 39–40, 43, 53–5, 59, 70, 88, 100–2, 104, 139–41, 176–80
locality 39–40, 43, 53, 70, 101–2, 104, 140–1, 144
   localities 9, 54–5, 59, 70, 117, 140, 177, 179–80
Longo, Giuseppe 102, 164 n.17, 187, 190 n.23, 190 n.28
Lotka, Alfred 3, 7, 36–8, 40, 45–6, 48 n.4, 48 n.6, 48 n.8, 109
loving memory 11, 205, 207–13, 213 n.2, 214 n.6, 214 n.16

malaise 5, 14 n.45, 34 n.7, 55, 59, 91 n.25, 105 n.12, 135 n.41, 150 n.40, 227 n.8
Mankind and Mother Earth (Toynbee) 35, 37, 48 nn.9–10, 48 nn.12–15
Marx, Karl 21, 23, 98, 122 n.29, 168, 194, 239
Mauss, Marcel 175, 180 nn.1–2
media 9–10, 17, 64, 146, 169–72, 184, 186, 189 n.5, 191–9, 201 nn.8–9, 201 n.18, 205
melancholy 2–4, 28–33, 119–20, 123 n.51
memory 2, 11, 24, 27, 29, 53, 55–6, 58–9, 64, 72, 97, 99, 101, 104, 127, 130, 133, 140–3, 147, 150 n.36, 155, 158, 180, 195, 205–13, 214 n.6, 218–20, 235
   mnemotechnics 5, 77, 81, 141, 205–7, 209–11, 213, 215 n.32, 228 n.28
Message Layer Security (MLS) 199
metaphysics 6–7, 22, 29, 41–3, 81–5, 87–9, 94 n.93, 98, 112, 127
milieu 5, 8, 17, 36, 46, 54, 56, 66, 79–80, 82, 87, 93 n.89, 94 n.89, 131, 140, 142, 146, 155, 161, 163 n.2, 169, 178, 185, 210
   associated milieu 80, 161, 165 n.37
mimesis 10, 193, 227 n.18
mnemotechnics 5, 77, 81, 141, 205–7, 209–11, 213, 215 n.32, 228 n.28
music 11, 157–8, 160, 217–26, 227 n.17, 228 n.38
   hearing aids 224–6
   instrument and recording 218–20
   jazz 217, 223, 225
   political 220–4
Musso, Pierre 192, 201 n.3

Naked Lunch (Burroughs) 19, 25 n.6
Nanjing Lectures 2016–2016 (Stiegler) 13 n.21, 99, 105 nn.28–30, 133 nn.1–2, 135 n.39, 151 n.71
National Security Agency (NSA) 199
Neganthropocene 12 n.4, 12 n.11, 13 n.14, 13 n.18, 23–4, 60 n.7, 61 n.9, 61 n.23, 74 n.42, 99, 104, 104 n.3, 105 n.19, 105 n.23, 105 nn.29–30, 117, 120 n.3, 121 n.15, 133 n.1, 152 n.94, 153 n.95, 173 n.5, 186, 189
Neganthropocene, The (Stiegler) 12 n.4, 12 n.11, 13 n.14, 13 n.18, 60 n.7, 61 n.9, 61 n.23, 74 n.42, 104 n.1, 104 n.3, 105 n.19, 105 n.23, 105 nn.29–30, 106 nn.37–9, 106 n.40, 133 n.1, 120 n.3, 121 n.15, 161 n.22, 133 n.1, 133 nn.3–4, 136 n.48, 151 n.61, 152–3 nn.94–5, 173 n.5
neganthropos 23–4
negentropic bifurcation 101, 103, 163
negotium 9, 155, 162, 168
neoliberalism 209
network architecture 162
network ideology 192, 201 n.3
networks 162, 192–3, 196–7, 200
   social networks 10, 139, 161–2, 168, 191–200, 209
NEXTLEAP 199
noēsis 5–6, 54, 64, 70, 81, 86, 89, 139, 141, 143, 146, 149 n.6, 171, 208
noēton 82
noodiversity 23, 70, 72, 75 n.54, 117, 177, 178
Nootechnics 20, 58, 61 n.17, 62 n.33, 105 n.18
normativity 66, 157, 162, 163 n.9, 168, 239
nous 4, 55, 80–1, 89

online generation 185
organology 8, 11, 22, 55, 58, 61 n.23, 103–4, 105 n.18, 117, 123 n.43, 145, 148, 155–6, 159–60, 199
   organological 6, 8, 23, 40, 56–7, 60, 64, 97, 103, 108–9, 112, 142, 144–6, 156–8, 163, 224

origin 4–5, 10, 27–30, 33, 80–1, 111, 141, 143, 191, 210
   original 21–2, 27, 30, 34 n.11, 66, 68, 77, 108, 130, 186, 192, 197, 211, 223, 226 n.1
   originary 4–7, 27–9, 53–4, 81, 109, 111, 121 n.16, 126–7, 129, 191, 217, 234, 237–8
otium 9, 155, 162, 168, 188

pansable (curable) 4, 53, 55, 60
panser 2, 7, 55, 112, 118, 121 n.16, 125, 128, 133 n.1, 145, 187
Parker, Charlie 110–11, 223
Pasquale, Frank 176–7, 181 n.8, 181 nn.10–11
pensable (that which can be thought) 4, 53, 55
penser-improviser (to think and improvize) 159–60, 164 n.22
Peters, John Durham 210, 214 n.23
Phaedo 87, 89
pharmakon 5, 19, 22–3, 31, 42, 47, 54, 60 n.1, 98, 115, 122 n.23, 142, 147–8, 161, 172, 185, 189 n.11, 190 n.18, 200, 205, 207, 210, 212, 214 n.12
   pharmacology 22, 42, 61 n.12, 89–90, 122 n.23, 155, 160, 195, 201 n.1
Phenomenology of Spirit (Hegel) 18, 31, 34 n.13
philia 160, 172, 194
philosophical relativity 12, 238, 239 n.3, 240 n.20
Philosophising by Accident (Stiegler) 13 n.34, 25 n.8, 150 nn.24–5, 214 nn.4–5, 214 nn.8–11, 214 n.14, 214 n.17, 214 nn.20–1, 215 nn.28–30, 215 nn.32–4
philosophy 1–3, 8, 11–12, 14 n.36, 17–18, 21–4, 39, 43–4, 53–5, 57–9, 61 n.11, 87, 89, 92 n.41, 104, 111, 126, 128, 140–4, 146, 191–2, 195–6, 206–8, 234, 236, 238, 239 n.3, 240 n.20
phonograph 11, 111, 206–7, 212, 220, 223, 228 n.28
Plaine Commune 9, 22, 147, 162, 164 n.25, 167, 169–71, 200
   Plaine Commune programme 9, 114, 147, 155, 167, 169

Plato 21, 23, 56, 81, 86, 104, 140–1, 146, 207–8, 212–13, 215 n.28, 220–2
Polemic Tweet device 161
politics 10, 21, 108, 112, 117–18, 126, 128, 147–8, 149 n.6, 170, 178, 180, 185–7, 199, 226, 227 n.18
practice(s) 8, 22–3, 55, 57–8, 67–70, 139–42, 144, 146, 149 n.6, 162–3, 180, 207, 223, 225–6
Prendre Soin (Stiegler) 185–6, 189 n.3, 239
presence 87, 112, 116, 127, 130, 139, 205–6, 209, 213, 237
present, the 55, 112, 114, 125–6, 128, 131–2, 134 n.20, 191, 205, 209–11, 224
primary retention 219–20
pro-gram 110, 112
programmatology 56
programme 157, 161, 163, 168, 169, 170
   programme of the improbable 223, 228 n.28
proletarianization 3, 9, 19, 22, 47, 68, 87, 147–8, 168–70, 181 n.13, 187, 199
Prometheus 28, 78, 111, 210
Protection Maternelle et Infantile (PMI) 171–2
psychopower 145, 197

reading 10–11, 18, 21, 36, 56–7, 64, 67–8, 80–1, 83, 85–7, 111, 118, 139, 145, 158–9, 211, 239
reason 19, 29, 31, 38, 42–3, 49 n.22, 57, 59, 64, 78, 83–6, 92 n.39, 113, 119, 132, 148, 159, 168, 194, 212, 237
retention 33, 77–8, 87, 91 n.33, 92 n.39, 94 n.94, 159, 180, 219–20, 223, 237
   retentional 4, 55, 80, 116, 133, 140, 142–3, 197, 220, 224
Rifkin, Jeremy 224
Riley, Denise 213, 215 n.35
Rilke, Rainer Maria 45–6, 49 n.32

Sahlins, Marshall 66–7, 73 n.13
Schrödinger, Erwin 36, 41, 44–6, 48 n.19, 49 n.30, 101–3, 105 n.24
scientific knowledge and politics 187
Scott, James C. 65–6, 68–70, 73 n.11, 74 n.38, 75 n.56

search for extra-terrestrial intelligence (SETI) 210
secondary retention 219–20
Seeing Like a State (Scott) 69, 74 nn.38–40
Sen, Amartya 168
sentiment funeste 30
Simondon, Gilbert 3, 10–12, 45–7, 49 n.33, 49 n.35, 54, 57, 59, 61 n.22, 62 n.37, 80, 102, 111, 142, 157–9, 161, 163 n.5, 165 n.32, 165 n.34, 165 n.37, 197, 210, 221, 233–6, 239 nn.2–3, 240 nn.4–6
singularity 5, 10, 14 n.37, 29, 63, 79, 87, 99, 101, 144–5, 149, 151 n.65
Sisto, Davide 209, 214 n.18
SixDegrees 192
Sloterdijk, Peter 18, 20–1, 24, 24–5 n.3, 25 n.14, 105 n.30, 237
Social Web 197–200, 201 n.18, 202 n.34
Socrates 87, 89, 144–6, 206, 208
sophist(s) 21, 143–5, 217
soul 5, 11, 22, 29, 36, 38–9, 42, 45–7, 68, 79–82, 85–6, 88–90, 91 n.18, 93 n.89, 94 n.89, 103–4, 141, 220–1
space 5, 45, 54, 57–8, 149 n.6, 158, 177, 179, 188, 196, 199, 210
speed 24, 58, 134 n.6, 168, 193, 209–11, 223
spirit 4, 8, 20, 31–2, 54–5, 60, 86–7, 99, 117, 139–44, 146, 155–9, 161–3, 206
Standard Generalised Markup Language (SGML) 157
States of Shock (Stiegler) 60 n.1, 60 n.3, 122 n.31, 149 n.10, 151 n.58, 179, 180 n.1, 181 n.6, 181 n.16, 181 nn.17–23, 182 n.31
stupidity 19, 23–4, 90 n.6, 168, 170, 186
subsistence 59, 67–8
suicide 2, 4, 39–40, 44, 53, 55, 58, 60, 63
supplement 54, 60, 86, 100, 109, 130, 200
supplementarity 130, 140
supplementation 127–8, 143–4, 147
support 11, 31, 47, 77–9, 114, 130, 139–41, 144, 148, 159, 171, 180, 185–6, 193, 198, 220
surpréhension 159, 164 n.21
suspensible (that which can be hung) 4, 53, 55

Symbolic Misery (Stiegler) 19, 220, 222, 224, 226, 227 n.4, 227 nn.15–17, 228 n.22, 228 n.32, 228 nn.34–6, 229 n.45, 229 n.49
synthesis 31, 41, 102–3, 158, 223, 239 n.2
Szendy, Peter 217, 223, 228 nn.25–6

tactical media 192, 195
technē 3, 62 n.39, 82, 236
technicity 4, 7–8, 87, 126–7, 129, 150 n.34
Technics and Time (Stiegler) 5–6, 11, 12 n.10, 18, 22, 27–8, 36–7, 78–9, 84–5, 98–9, 107–11, 113–14, 116, 120 n.2, 126–7, 145, 195, 218, 220, 235
Technics and Time, 1 (Stiegler) 18, 37, 98, 116, 126–7, 218
Technics and Time, 2 (Stiegler) 99, 195
Technics and Time, 3 (Stiegler) 84–5, 99, 220
technologies of spirit 155, 158, 163 n.4, 185
technoscience 61 n.24, 82, 92 n.39, 148
tekhnē 111–12, 131, 134 n.10
television 156, 181 n.9, 223
temporality 35, 58, 84, 87–8, 113, 117, 120 n.2, 127–8, 130, 155, 211, 213, 219–20
tendency/tendencies 3, 19, 38–43, 47, 54, 56–7, 63, 72, 81, 101–2, 126, 132, 175, 177
textuality 14 n.37, 158, 209
thermodynamics 3, 37, 39–40, 43–4, 97–8, 101–3, 188
Thiel, Peter 10, 193–4
Thunberg generation 156, 183–9
Thunberg, Greta 20, 132, 183, 186
time-form of environmentality 116
time/temporalization 6–7, 11, 23, 36–7, 77–80, 97, 100–1, 103–4, 108–11, 113–16, 118–20, 126–8, 130–2, 145, 220–1
Toynbee, Arnold 3, 35, 37–9, 41, 46, 48 nn.9–10, 48 nn.12–15
transhumanism 12 n.10, 30, 185
transindividuation 3–4, 10, 45–6, 54, 57, 113, 115, 142–4, 150 n.36, 157–9, 161, 179, 197–200, 214 n.9

Tristes Tropiques (Lévi-Strauss) 13 n.20, 100, 105 nn.20–1
truth 12 n.10, 17, 19, 21, 23, 61 n.21, 64, 81, 84–5, 108, 113, 120, 130, 183, 187, 193

Umbrella generation 184–5
Uncontrollable Societies of Disaffected Individuals (Stiegler) 61 n.15, 92 n.47
Université de technologie de Compiègne (UTC) 156

value 3, 29, 41–3, 46–7, 58–9, 65, 68, 163, 172, 187, 195, 209, 239
vécu/lived experience 83
Vernadsky, Vladimir I. 3, 35–6, 48 n.1, 48 n.7

What Makes Life Worth Living (Stiegler) 61 n.12, 61 n.32, 121–2 n.23
will, the 19, 29, 88
Wolf, Maryanne 195
work 6, 9–10, 67–68, 97, 155, 168
work of art 10, 223, 225
World Trade Organization (WTO) 192
World Wide Web Consortium (W 3C) 10, 163 n.9, 198

XMPP. *See* Extensible Messaging and Presence Protocol (XMPP)

Zuckerberg, Mark 192–3

www.ingramcontent.com/pod-product-compliance
Lightning Source LLC
Chambersburg PA
CBHW070723020526
44116CB00031B/1392